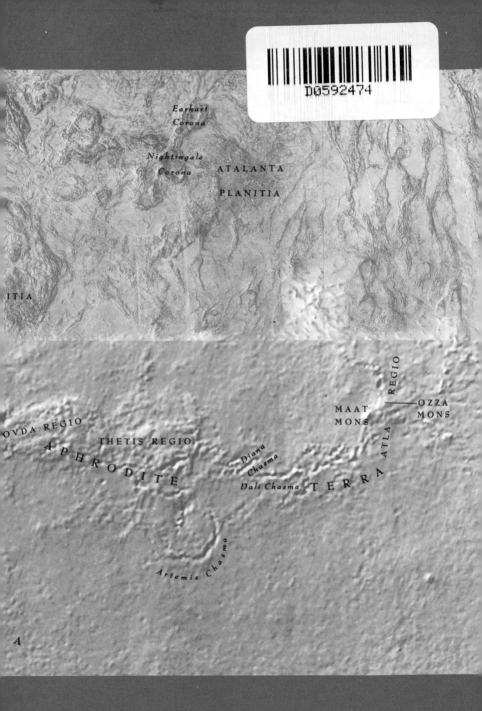

Earhart
Corona

Nightingale
Corona

ATALANTA

PLANITIA

ITIA

REGIO

MAAT ——OZZA
MONS MONS

OVDA REGIO

THETIS REGIO

APHRODITE

Diana
Chasma

Dali Chasma T E R R A

ATLA

Artemis Chasma

A

MAP OF VENUS

Of all the planets in the solar system, none has captivated the imagination of man more than Venus. As early as 3100 B.C., Venus was identified with the goddess of love, and until the 1960s it was believed that it might actually harbor life. Edgar Rice Burroughs populated Venus with a beautiful princess, a cast of dastardly villains, and of course a dashing hero. But the science of Venus has only recently been understood and it tells a very different story—of a planet with a surface temperature of 900° Fahrenheit, a surface pressure ninety times that of Earth's at sea level, and a carbon dioxide atmosphere that is the prototype for all our fears of the greenhouse effect.

In *The Evening Star*, Henry S.F. Cooper, Jr., veteran science and space reporter for *The New Yorker*, tracks the Magellan spacecraft that has been mapping Venus from orbit since August 1980. In eloquent, vivid prose, Cooper introduces us to the engineers who have nursed the spacecraft's fragile electronics and the scientists who have used the spacecraft's data to assemble a picture of this strange new world.

In size, density, and composition, Venus is almost identical to Earth, yet its nature and history turn out to be as different as close relatives frequently can be. Why did Venus develop a hot, heavy atmosphere, while Earth did not? Why did Venus take one evolutionary pathway and Earth another? As the scientists wrestle with these and other questions, the engineers play tricks on the spacecraft's balky computer to keep the scientists supplied with the data they need.

With a keen eye for irony and a knack for making the most sophisticated technology eminently accessible, Cooper high-

(continued on back flap)

THE EVENING STAR

Venus Observed

BY HENRY S.F. COOPER, JR.

HENRY S.F. COOPER, JR.

THE EVENING STAR

Venus Observed

FARRAR STRAUS GIROUX

NEW YORK

LIBRARY OF CONGRESS CATALOGING-IN-PUBLICATION DATA
Cooper, Henry S. F.
The Evening Star : Venus observed / Henry S.F. Cooper, Jr.—1st
ed.
p. cm.
1. Venus (Planet)—Popular works. 2. Magellan (Spacecraft)—
Popular works. I. Title.
QB621.C72   1993   523.4'2—dc20   92-36347 CIP

The front endpaper is a map made prior to Magellan's
arrival at Venus. It combines altimetry measurements
from the American Pioneer 12 spacecraft (launched in
1978) with a resolution of about 150 kilometers and
airbrushed cartography based on radar imagery of the
northern third of the planet from the Soviet Venera 15
and 16 spacecraft (launched in 1983) with a resolution
between one and three kilometers.

The back endpaper is a map made more than two
years after Magellan's arrival at Venus in August 1990.
It combines radar imagery from Magellan's first three
cycles with a resolution between 120 and 300 meters—
about a tenfold improvement over the Soviet Veneras.
The gaps represent data lost due to superior conjunction,
"hide" maneuvers to prevent spacecraft overheating, and
other problems.

Both maps courtesy of the U.S. Geological Survey,
Flagstaff, Arizona

In Memory of
William Shawn

CONTENTS

PREFACE

This book is dedicated to the memory of William Shawn, who died as it was going to press in December 1992 and who from 1952 to 1987 was editor of *The New Yorker*. During his tenure, he created an environment where writers could prosper. He elicited their best efforts in a variety of subtle ways, among them by encouraging writers to develop their own fields of interest, on the theory that they would write better about a subject they had picked than about one they were assigned. I myself, having arrived at the magazine not long after Sputnik, in the next decade started turning out stories about the exploration of space, and I have continued to do so ever since.

Many writers on *The New Yorker*'s staff of my generation—we are now in our late fifties or early sixties—seemed naturally to gravitate toward scientific subjects, even though almost all, myself among them, had majored in the arts (particularly English literature) in college. Over the years, they have included Jeremy Bernstein, Paul Brodeur, Gerald Jonas, Horace Judson, John McPhee, and William Wertenbaker. (Bernstein, a physicist, is an exception to the liberal-arts background of these writers.) It was

....

greatly to Mr. Shawn's credit that, although he personally was not much interested in science, he nonetheless gave us a free hand to turn out long articles about science and its practitioners, and (sometimes in spite of himself) he usually published what we wrote.

It is not hard to see why so many liberal-arts majors of this generation have wanted to write about science. For one thing, of course, in the nineteen-fifties and -sixties—our formative years as writers—science was making great strides, particularly in nuclear physics, molecular biology, geology, and now the study of the other planets in the solar system. Science, in other words, was very much where the action was.

Furthermore, many of us from the liberal arts may have been influenced, consciously or unconsciously, by the ideas of Lord Snow, who argued in his book *The Two Cultures and the Scientific Revolution* (1959) that science and the arts, which throughout most of history had been very close, had drifted apart in the last century and a half, largely because of the increased specialization of science, so that the scientists could scarcely be understood by their erstwhile colleagues. For me it became a challenge to help bridge the gap—to see whether I could explain to other nonscientists what scientists were doing, how they worked, what they thought, and maybe even what they were like. In this I (and I believe many others) was abetted by a conviction that people in the arts and people in the sciences were, in fact, going about the same business in their different ways—the business of explaining and interpreting the world and perhaps also what lay outside it—and that therefore they were not

....

so different from one another after all. Evidence for this theory is that in addition to the hundreds of writers who over the last thirty years have embarked from the arts side of the gap, a number of highly literate scientists—among them Lewis Thomas, Carl Sagan, and Stephen Jay Gould—have set out in our direction from the far shore, with brilliant success.

With respect to William Shawn, I suspect that it was our attempts to portray science as a humanistic endeavor, carried out by people not altogether different from his writers, that led him to print our work in his magazine. One aspect of *The New Yorker* under Mr. Shawn was that he encouraged writers to throw themselves into a new experience as though it were a river (in our case, it was usually one running through Lord Snow's gap), and follow it wherever it went. This adventurous approach lent itself particularly well to writing about the exploration of space. Usually, when I set out to report a planetary mission, I would flounder along without the foggiest idea of where the current was taking me. I was invariably rescued by the scientists, who turned out to be extremely good at explaining to me in excellent English what they were about—it may be, in fact, that Lord Snow's gap was not all that broad to begin with. Usually, the more accomplished these scientists were, the more they enjoyed the challenge of interpreting their research to me. All I had to do was write down what they said.

I have found this to be true regarding all the ports of call I have made across the solar system, beginning with the moon, and following our fleet of spacecraft (and sometimes the Soviet or Russian fleet) to Mars and its moon Phobos, to Saturn, Uranus, and Neptune, and their moons. Gathering information for a book about the Magellan mission to Venus was no exception. Many scientists and engineers are responsible for hauling me out of the water, drying me off, and setting me in the right direction. For starters, I should thank everyone mentioned in this book, who at one time or another shared an insight or a bit of excitement with me. In particular, I should single out David Okerson, who, along with John P. Slonski, Jr., guided me through the intricacies of Magellan's almost fatal computer glitches; and also Alexander T. Basilevsky, James W. Head III, Roger J. Phillips, R. Stephen Saunders, Gerald Schaber, Sean C. Solomon, and Anthony J. Spear, who made their scientific or engineering work—and disputes—come alive for me.

Many of the above read the manuscript, as did Jayne C. Aubele, Ellen R. Stofan, and Richard Vorder Bruegge. Many others read parts of it, including Gerald Schubert and Laurence A. Soderblom. In addition, Gordon H. Pettengill, Raymond Batson, Peter Ford, and Randolph L. Kirk, among others, answered checking questions on the telephone. Haig Morgan and Robert Sucharski helped with the maps. I am grateful to all of them. Though they provide a formidable fence against error, they cannot prevent a former English major from getting something wrong in science or technology, and they are not responsible for any errors in this book. Any remaining mistakes are evidence that Lord Snow's gap still exists, though I hope after all these years, and with so many efforts by so many scientists and writers, it is getting smaller and smaller.

Henry S.F. Cooper, Jr.
New York, New York
December 1992

PART I

The Evening Star

THE Magellan mission currently in orbit around Venus is close to the last space flight of the classic era of planetary exploration, when unmanned spacecraft —Pioneers and Mariners, Vikings and Voyagers— of exquisite delicacy and complexity, crammed with instruments, returned data, and particularly pictures, from unknown worlds, to teams of excited scientists waiting anxiously at the Jet Propulsion Laboratory in Pasadena. The Mars Observer mission which was launched in September of 1992, the Galileo mission currently bound for Jupiter, and the Cassini mission planned for Saturn are part of this tradition; however, Galileo and Cassini are in trouble, one for mechanical reasons and the other for financial ones. The next generation of missions will be smaller, specializing in one or another aspect of physics or chemistry. There will be fewer pictures, if there are any at all, and increasingly the scientists will be receiving their data at their own laboratories, instead of clustering excitedly around computer terminals at J.P.L.

VENUS has beckoned since the dawn of time. Seen from Earth, the planet is a tiny brilliant disk smaller than a lowercase o in a book held at arm's length, so small it may not even appear to be a disk except to the sharpest pair of eyes. Seen through a pair of binoculars, it can also be a crescent, like a parenthesis, open or closed, for Venus, being nearer the sun than Earth is, has phases like the moon. It is yellow-white, more white than yellow; its cloud-laden atmosphere (largely carbon dioxide but with some sulphur) is so reflective that Venus is five times brighter than the full moon and so opaque that the surface cannot be observed visually. It is so bright that Venus, like the moon, can be seen in daylight, up to three hours before sunset or after sunrise; because of its brilliance, we call it the Evening Star or, alternately, the Morning Star. In the evening, when most people have the time to contemplate it, Venus can play a strong supporting role in a sunset, its hard white light gradually gaining in intensity as the orange and saffrons, pinks, and reds fade; then, the sun gone, it dominates the western sky until the flood of stars encroaching from the darkened east overtakes and surrounds it, leaving it still a commanding presence of light—dimmed only by the moon, a far larger bulb but of lower wattage.

It has long beguiled mankind. One NASA official, Wesley Huntress, the director of solar-system exploration, believes that if it weren't for the beguiling presence of the moon, man might never have thought to leave Earth; however, minus the moon, Venus might have provided an enticing substitute. The moon aside, Venus is our nearest neighbor, the planet next in to the sun, from which its orbit is 67,200,000 miles distance; our orbit is 92,957,200 miles out. In the course of its year, which takes 225 of our days,

Venus passes within 26 million miles of Earth—closer than Mars, our next nearest neighbor, which approaches no closer than 35 million miles. As Venus is almost exactly the same size as Earth (7,503 miles in diameter, as opposed to 7,906 miles in diameter) and almost exactly as dense (5.3 grams per cubic centimeter, as opposed to 5.5 grams per cubic centimeter), it was long thought to have been Earth's twin. It was believed, up until the 1960s, that Venus might be Earth-like with respect to harboring life, and writers of science fiction endowed its surface with advanced civilizations—such as Edgar Rice Burroughs, who populated Venus with a beautiful princess, all sorts of dastardly villains, and a dashing hero—all shrouded from our eyes by the clouds.

Venus has been visited by spacecraft more often than any other planet; only the moon has been visited more frequently. Beginning with our Mariner 2 mission launched in 1962, six American spacecraft and eighteen Soviet ones have flown by, orbited, or landed on Venus. Collectively, they have shown that Venus is about as hospitable to life as a pressure cooker. The temperature at the surface—900° Fahrenheit— is hot enough to melt lead, zinc, and tin, and the surface pressure is ninety times Earth's at sea level, or the same as it is 2,700 feet deep in the ocean. If there ever had been a princess on Venus, she would have died instantly, and a medical examiner would have had a hard time deciding whether she had been fried, crushed, or suffocated, or all three, in the searing, heavy, carbon-dioxide atmosphere—the prototype for all our current fears of the greenhouse effect. Perhaps the most dramatic result of our planetary program has been to prove, one by one, that all the solar system's planets, other than our own, are inhospitable to life and barren. And Venus has provided the additional bad news that we are in danger of following her deadly example and ourselves becoming a planetary autoclave.

Fortunately, planetary exploration had another major thrust aside from a search for life—the gathering of chemical, physical, and geological information about the planets and their moons, with an eye to placing Earth more firmly in the context of the larger family of the solar system. Thirty years ago, we knew well only one planet, our own, which was like having a single example of a species. Today we are conversant with seven more planets and fifty-odd moons; we are beginning to understand the history and relationships of the family of planetary bodies and why different ones have evolved differently. We are beginning to understand the alternative paths Earth might have followed. But the most intriguing question in the field that has come to be called comparative planetology is why two such close siblings as Earth and Venus should have evolved in such different ways.

ON August 10, 1990, the Magellan spacecraft was approaching Venus's north pole at 24,600 miles per hour—so fast that if it didn't slow itself down, it would keep right on going, like an overeager bee overshooting a flower. Magellan had been launched aboard the space shuttle on May 4, 1989, a year late due to the Challenger disaster in January 1986.

Because of the relative positions of the two planets, Magellan had followed an unusually long trajectory to Venus, orbiting the sun one and a half times as it wound inward to its rendezvous. As it got closer to the planet, Venus grew to be an uppercase O (or, more precisely, a capital C, as it was only partially illuminated), then a whitish crescent of, say, a tennis ball palely imitating the sun, and finally a dazzling luminous crescent filling most of the sky. The thick whitish clouds reflected light brilliantly—the reason it can be seen from Earth in daylight. Unlike all the other terrestrial planets and all the moons (with the exception of Saturn's methane-enshrouded moon Titan), Venus's surface cannot be photographed or televised; it can only be sensed by radar, whose pulses can be assembled into images.

As most unmanned spacecraft do, Magellan looked like a surrealistic bug, sculpted perhaps by Joan Miró—not a bee, exactly, or a mosquito, or a water beetle, but a little like all three. It was rather squat and inelegant, compared to the graceful Voyagers with their long filigreed booms, true dragonflies of space. Its head is a dish antenna 3.7 meters across—this is the main antenna used both for transmitting and receiving data and for transmitting and receiving radar signals. A long thin cone that rakes backward from the lip of the big dish resembles an insect's feeler bent out of shape—a separate antenna for a part of Magellan's radar sensor that makes altimeter measurements. Its upper body is a rectangular box, the forward equipment module, which contains the electronics for the spacecraft's single scientific instrument, the synthetic-aperture radar, or SAR, which would map Venus through its clouds. Its lower body is an octagonal drum full of batteries, two computers, and thrusters for attitude control. On either side is a stubby solar panel—unusually small in proportion to the spacecraft because of the intensity of the sun near Venus, which meant that a smaller area of photovoltaic cells was needed than farther out in the solar system; the panels looked less like dragonfly wings than like water-beetle paddle feet. The entire craft, twenty-one feet long, was painted white or covered with white insulation blankets, a new mix of materials called Astroquartz which included glassy fibers of the Beta cloth used in the astronauts' space suits to protect them from sunlight. In addition, much of the spacecraft was covered with mosaics of square frosted-glass mirrors lest the heat be reflected with great intensity onto delicate parts of the craft.

At the base of Magellan was a detachable rocket, the solid-rocket booster, which would fire to insert it into orbit on August 10, at 9:42 a.m. Pacific daylight time, when Magellan—its trajectory bent by Venus's gravity—had passed behind the planet as seen from Earth. It would slow the spacecraft from 24,600 miles per hour to 18,675 miles per hour, enough for Magellan to be trapped by Venus's gravitational field.

At 9:15 that morning, I was sitting in a cubicle in the Magellan project office at J.P.L. with Keith Hamlyn, the propulsion system lead engineer. J.P.L. is an attractive jumble of buildings, fountains, and courtyards cascading down the side of a foothill of the San Gabriels behind Pasadena. But as far as the Magellan engineers and scientists were concerned, they might as well have been at the bottom of a cave on the dark side of Venus, for their quarters, on the second floor of a large low building named the Space Flight Operations Facility and called phonetically the SFOF, were without windows. The engineers were in a large loft of a room, painted gray, with gray carpeting. It was divided into dozens of small cubicles, the offices of the engineers for the various spacecraft systems, such as the computers, electrical power, and thermal control. The entire room—in its divisions a sort of analogue of the spacecraft—looked less like a mission-control center than like a floor of offices at a bank, and indeed it was about as lively. The spacecraft wasn't controlled in any direct sense. Rather, commands on tapes were uplinked every couple of weeks if nothing much was going on (or every couple of days if something was) to the onboard computers, which carried them out at the assigned times.

A more typical control room was on the floor below, where technicians were hunched gnome-like over consoles twinkling with screens and computers. They were connected with the Deep Space Network, whose global array of antennas in Madrid in Spain, Canberra in Australia, and Goldstone in California transmitted the command tapes to, and kept track of, not only Magellan but the rest of J.P.L.'s planetary fleet, including the Jupiter-bound Galileo; the two Voyager spacecraft, their tour of the outer planets behind them, now on their way to the outer edges of the solar system and beyond; and Pioneer 12, which was launched to Venus in 1978 but was still orbiting quietly, its mission complete—it would plunge into the Venusian atmosphere in October 1992. (One more craft would be added to the fleet in a couple of months—Ulysses, bound for Jupiter, which would swing it over the top of the solar system into polar orbit around the sun.) Each spacecraft had its own flight controller, called an Ace, who handled its communications. (Ace is a mutated acronym for assistant chief of mission operations, which became A.C., and then Ace; the Ace's backup quickly became known as the Deuce, thus confusing the etymology altogether.)

When Magellan disappeared behind the planet at 9:40, it could no longer be communicated with—a situation space engineers call LOS, for loss of signal. The instructions for the burn, called Venus orbit insertion, or VOI, had already been sent up to the spacecraft's computer—or uplinked, as space technicians say—and now there was nothing for Hamlyn to do but watch and wait. No one would know if the burn had been successful until after the spacecraft came out from behind the planet and it was in radio contact again.

Hamlyn is a large Englishman in his early thirties with a bushy brown beard who is normally quite placid but right now was twitchy. He lives in Denver, where he is employed by the Martin Marietta Aerospace Corporation, the prime contractor for Magellan. Most of the flight controllers are Martin Marietta employees—they are based in Denver, but many of the key ones spend much of their time at J.P.L., where the flight director is and the key decisions are made. Those who stay in Denver are in frequent contact with their counterparts at J.P.L. over a squawky intercom.

"I was perfectly all right until LOS," Hamlyn said.

"Now I'm getting worried." He clutched his TV monitor like a security blanket. "A watched pot never boils, I know, but I'm still going to watch it." He turned his attention to a second, larger monitor, full of numbers—the last data readouts from the spacecraft before it disappeared behind Venus. Normally, the numbers would change, reflecting the spacecraft's metabolism—increases and decreases in temperatures or pressures or electrical currents—but now they were still. With a thick grease pencil, he marked the spot on the screen that would be the first indication of a successful burn, the fuel-tank pressure readings, which should show up at AOS, or acquisition of signal, when the spacecraft emerged from behind the planet at 10:06, now nine minutes away.

The spacecraft had had an uneventful trip from Earth. There had been some minor problems. It had been overheating for reasons that were not clear, and this meant that it had to be oriented so that its round, dish-like radio antenna could be used as a parasol. Then the star scanners—meant to test the spacecraft's guidance system by finding certain stars—failed a number of its star checks. And then one of the gyroscopes used in navigation went bad and had to be turned off. (There are four of them, so there was plenty of redundancy.) There was a difficulty with one of the computers, too. All these problems had been solved, or gotten around—they were, in fact, not unlike the problems that come up on any space-flight. Although the flight controllers had every reason to believe the insertion burn would work, a little apprehension seemed a useful precaution—like knocking on wood. At a press conference the day before, the project manager, Tony Spear, a stocky, graying man with a small graying mustache who seemed more relaxed than the crisply businesslike engineers on either side of him, had remarked that if the rocket didn't fire, Magellan would become a flyby mission instead of a Venus orbiter; it would go into orbit around the sun, and it wouldn't be in a position for another try at Venus for a hundred years. There was nervous laughter. Although such a mishap had never happened to an American planetary mission, it had happened a few times to early Soviet spacecraft attempting to orbit Venus and Mars.

Indeed, if it hadn't been for some good luck and quick thinking before Magellan's launch, its rocket would not have worked. While Hamlyn stared at the still TV monitors, he told me that an engineer from the Morton Thiokol Company which had made Magellan's rocket, Keith Phillips, had gone down to Cape Canaveral to arm it. Another engineer read him, step by step, the instructions. After he returned to Baltimore, where the Thiokol factory was located, he woke up in the middle of the night with a sick feeling that he had hooked up the wires wrong. He was right. "What happened was that there had been a miscommunication on the procedures, between Phillips sitting under the rocket motor and the other guy reading the instructions ten feet away," Hamlyn told me. "One nut had a lock wire on it and the instructions were to remove the other nut—the one without the lock wire. One was black and one was shiny. The one to remove was the shiny one, but it, too, had a bit of wire on it that looked like a lock wire, so Phillips had removed the other one instead. I had never sat down with these guys and said, 'Not this one, but that one.' "

I asked Hamlyn what conceivably could go wrong now. "The rocket might fail to ignite," he said. "But not because the wires are hooked up wrong." He glued his face back to the screen.

There was less than five minutes until the spacecraft should reappear, assuming the burn had gone correctly. A dicey moment was approaching right now —the moment when the spacecraft would reappear if the rocket had not fired to reduce velocity. No one was seriously expecting the signal then—but it was still a time for fretting. An engineer from the next corridor stuck his head over the partition and said, "Fifteen seconds to 'no signal,' Keith"—"no signal" being what they hoped they would get.

"I'm not looking for it," Hamlyn said, waving him away. The moment came and went, uneventfully.

The same head reappeared over the partition. "That means it burned," he said.

"Of course it did!" Hamlyn said, with a trace of testiness.

The cubicle was gradually filling with well-wishers.

"Three minutes to go," Hamlyn said, slapping the knee of an engineer who had sat down on the chair next to him.

Over a squawk box, we could hear the project manager say to one of the DSN Aces downstairs, "We'd appreciate hearing from you as soon as we hear a signal."

"Yes, indeed we would," said Hamlyn, clutching his desk as if it were the control panel of a spaceship about to go into battle.

With thirty seconds to go, Hamlyn said to the

engineers filling his cubicle, "O.K., here we go, guys, stand back!" More engineers crowded in.

"This is the hot cubicle," one of them said.

"He's earning the big bucks we pay him today," another said.

The number above the grease-penciled line began to twitch and transform itself into another lower number—the first sign that the signal had been acquired and that fuel had been expended. Hamlyn leapt to his feet and gave a rebel yell—with a British accent. Everyone clapped him on the back. "Holy shit!" he said. One of his friends told me to make sure I wrote down "Holy cow." Either way, however often spacecraft have been inserted into orbit before, or how many times its trajectory is tweaked with mid-course corrections, or however many computers may have been involved, ringing a planet with a spacecraft always provides an undeniable kick—like throwing a horseshoe and watching it skirl around a distant post.

The screen was coming to life now, as more and more new numbers replaced old ones, reflecting the changed status of other systems. Rebel yells and sounds of backslapping began coming from the other cubicles. Two cubicles away, where engineers were monitoring the electrical current, a jubilant voice shouted, "Batteries charged, Captain Power!" In their sanitized, high-tech offices, flight controllers like to keep in touch with the science-fiction roots of space flight. To alleviate the gray monotony of their cubicles, some of the controllers post articles about space from supermarket tabloids—such as one, to the left of a cubicle door, that was headlined, "First Ever Photo of Captured UFO," over a picture of a disk-shaped spaceship hovering in a Soviet hangar; it looked like a frame from the film *Close Encounters of the Third Kind*, and probably was.

There was some mystification in the guidance and navigation cubicle, though, for the spacecraft had apparently switched for no known reason from its primary to one of its backup gyroscopes, which are mechanical devices that always know their orientation in relation to the stars, as if they were compasses for space: they operate on the principle of a child's top, though with lots of electronics. Very likely, the engineers thought, the burn had caused the spacecraft to oscillate slightly, and the easiest way the guidance computer could cope with the situation was to assume that the space compass, not the spacecraft, was at fault.

A few minutes later, the visiting engineers had returned to their cubicles, and once more all was silent as a bank vault. The spacecraft settled into its highly elliptical orbit—its periapsis, or point of lowest approach, was 265 kilometers above the backside of the planet; but its apoapsis, or most distant point, was 8,427 kilometers above the surface. The spacecraft took three hours and twenty-six minutes to make a complete circuit, which it did 7.3 times a day.

That night, at apoapsis of the fourth orbit, some explosive bolts fired to separate the spent rocket motor from the spacecraft. Instantly, there was another unexplained swap of gyros—from the backup back to the prime. The spacecraft apparently rocked back and forth as a result of the blast, and apparently once again the computer had assumed that the space compass, not the spacecraft, was at fault. Seven sec-

....

onds later, several alarm signals indicated that there were some misaddressed commands in the backup computer and its memory, memory B. In the memory there were 32,000 sort of mailboxes, each of which contains what is called a word but is actually a string of sixteen binary numbers; the words, in turn, are codes for instructions which, when they are called up by the spacecraft's computer, can be interpreted and turned into actions. (All computer words are made of strings of various combinations of ones and zeros, depending on whether a switch—within a silicon chip, the basic building block of a computer—is on or off.) Another part of the computer, the central processing unit, or processor, knows when a certain action needs to be performed, and consequently sends a signal to the mailbox that contains the right instructions. Each mailbox has its own address, which is four bits long, and it appeared that the fourth bit in the address logic of four of the addresses was stuck at the binary number one, so that, regardless of whether the last digit in these addresses was zero or one, it read one. The alarms in the backup computer continued intermittently for a day and a half.

The problem did not affect operations because the backup computer was not on line; the prime computer was running the spacecraft. Had the backup computer been serving as the prime, the spacecraft might have carried out some wrong actions and gotten itself into trouble before the fault-protection system was alerted and swapped computers—it might even have resulted in something called a runaway program execution and something else called a heartbeat failure, scary events flight controllers don't like to think about.

....

They didn't know what the problem was, though they suspected it lay in the computer's hardware—very likely a damaged silicon chip—as opposed to its software. (The hardware is the physical machinery; the software is the program that makes it do whatever it does.) The controllers arranged matters so that memory B was made inaccessible to the primary computer; normally, the primary computer kept memory B up to date—the latest navigation information is constantly being transmitted from the ground—but now memory B was no longer being updated. However, memory B was kept running and was continually monitored by the prime computer to see if the problem would recur. Though no damage had been done, the flight controllers were uneasy; they didn't like unexplained events.

LONG before Magellan arrived at Venus, people wondered about it. As early as 3100 B.C., it was identified with the goddess of love. About 1610, Galileo Galilei with his telescope was the first to notice the phases of Venus, from which he deduced that it orbited the sun—a discovery that heretically undermined the Earth-centered universe. In the latter part of the seventeenth century, Christian Huygens, the Dutch astronomer, suggested that Venus had an atmosphere. "I have often wondered that when I have viewed Venus . . . she always appeared to me all over equally lucid, that I can't say I observed so much as one spot on her," he wrote (as quoted in the article on Venus, written by Carl Sagan, in the Encyclopaedia Britannica). "Is not all that light we see reflected from an atmosphere surrounding Venus?" In 1761, just after Venus crossed in front of the sun's disk in what is called a transit, Mikhail Vasilyevich Lomonosov, a Russian astronomer and physicist, proved not only that Venus had an atmosphere but that it was at least as thick as Earth's. Through his telescope, he noticed an arc of light surrounding part of Venus's otherwise dark disk; the phenomenon could only be caused by the refraction of sunlight in an envelope of gas. (Although Venus passes between Earth and the sun roughly every eighteen months, transits of Venus are rare because the three bodies are seldom precisely enough aligned in the same plane.) At first, astronomers believed the atmosphere was relatively clear, like ours; and indeed in the 1770s the English astronomer Sir William Herschel, who confirmed the existence of the atmosphere, thought that through it he could see spots on the planet's surface. He was wrong. In 1918, the Swedish astronomer Svante August Arrhenius, who thought he detected clouds in the atmosphere, decided that the clouds meant water, and suggested that Venus was a swampy place redolent with low forms of plant life. He was wrong. In the 1920s two other astronomers, Charles Edward St. John and Seth Barnes Nicholson, unable to detect water in a spectroscopic analysis of Venus's light, suggested that Venus was a dry, sandy planet whose clouds were made of windblown dust —as happens on Mars. They were wrong. Nonetheless, Arrhenius, St. John, and Nicholson were part of the scientific substrata that allowed Burroughs a few years later to locate his princess on a planet with oceans and continents, forests and deserts, not unlike our own. He was wrong.

No reliable information was gathered about the surface of Venus until the mid-nineteen-fifties, when radio astronomers trained their big dish antennas on Venus and began listening for its natural radio emissions such as all bodies, including you and I, constantly radiate. They suggested that the surface temperature was approximately 900° Fahrenheit—something that the astronomers, in the days before the runaway greenhouse theory for Venus had been worked out, were reluctant to credit. Subliminally, they may have been unwilling to give up the lure of Burroughs's lush, watery Venus with its exotic royals. The astronomers searched for other explanations for the high emission from Venus—some postulated intense lightning, others a dense ionosphere—and it wasn't until 1964 that a Soviet scientist and an American scientist, A. D. Kuzmin and B. G. Clark, using the California Institute of Technology's antenna at Owens Valley, California, proved that the high temperature readings could come only from the surface. Exit the princess.

Aside from natural radio emissions, radio astronomers had another tool at their disposal—radar. Radio waves transmitted toward a hard object, as British scientists had learned just before World War II, would bounce back, penetrating any intervening clouds; with the proper equipment, the signals could be received by the same antenna that had transmitted them. The first waves bounced off another planetary body occurred in 1946, when scientists at the U.S. Naval Research Laboratory zapped the moon. Because it was so far away, Venus was a tougher target, and like much else in the study of Venus, credit for the first success is hazy. The first attempt, using an antenna at Millston Hill, Massachusetts, was made in 1958 by a group of scientists at the Massachusetts Institute of Technology, which included a young radar physicist named Gordon H. Pettengill, who several decades later would be the principal investigator for the radar instrument aboard Magellan. "We had all sorts of new equipment which had just become available—a maser, which amplifies a microwave beam in the same way that a laser amplifies a light beam, and digital data-processing techniques," Pettengill told me. "Suddenly the technology for doing this was coming together." The M.I.T. scientists thought they received an echo from a hard surface, and published the finding in journals. However, in 1959, when they tried a second time, they thought they found no echo. In 1961, several groups in this country, a group in England, and a group in Russia, tried again. This time, a young graduate student at the California Institute of Technology, Richard M. Goldstein, using an antenna at Goldstone, got an unmistakable echo. When the M.I.T. group went over its earlier data in light of the new readings, they discovered that in 1958, when they thought they had received an echo, they had not; and in 1959, when they thought they hadn't, they had. "But we didn't know that until 1961," Pettengill said. "So I guess you would have to say that Goldstein in 1961 gets the credit." Information about the surface of Venus would continue to be elusive. Whoever sent and received back the first signals from Venus, the scientists had proved that underneath the clouds Venus was a solid body; moreover, the radar reflectivity of the surface, fifteen percent, suggested that the planet's surface was rock. (The moon, covered with a deep layer of dust, has a radar reflectivity of only seven percent.)

As equipment improved, the radar study of Venus began to heat up. Cornell University's big antenna at Arecibo, Puerto Rico, and Cal Tech's and J.P.L.'s antennas at Goldstone soon turned their attention to Venus, too. Though the first images were very crude, the scientists succeeded in identifying a few major features on the surface, and by timing their disappearance and reappearance with the planet's rotation, the scientists discovered that Venus rotates from east to west (the opposite way from all other planets except Uranus and from almost all moons) and that a Venus sidereal day, 243 Earth days, lasts longer than a Venus year, 225 Earth days. Indeed, a Venus day coincides almost exactly with the interval between the closest approaches of Venus to Earth, which means that Earth-based radar always gets its best look at the same side of the planet.

In December 1962, the first of a long line of American and Soviet spacecraft, NASA's Mariner 2, flew by Venus, passing about 21,000 miles from the planet and establishing that it had no magnetic field —a fact which some astronomers thought might be related to the planet's slow rotation. Like most of Mariner 2's successors, it carried no radar. In 1967, the Soviet Venera 4 dropped a probe with a parachute into the atmosphere which sent back data on its composition for ninety-three minutes, until it ceased operating; Venera 4 and our Mariner 5 flyby mission, also in 1967, confirmed the atmosphere's high carbon-dioxide content—ninety-seven percent—with only small traces of nitrogen, sulphur dioxide, argon-40, argon-36, oxygen, carbon monoxide, and water vapor. The high percentage of carbon dioxide, of course, is responsible for the so-called greenhouse effect, because the sun's visible light passes easily through it to the surface, but the heat radiated from the surface as infrared light cannot pass out through the gas. (Even the limited amount of carbon dioxide in Earth's atmosphere keeps the temperature twenty-five degrees warmer than it would be otherwise—a figure that appears to be rising because of our consumption of fossil fuels.) Several Soviet Venera spacecraft which attempted landings failed because of the great heat and pressure lower in the atmosphere—though they succeeded in returning data about the atmosphere.

Venera 7 in 1970 was the first craft to land softly on the surface and send back information; it transmitted data (but no pictures) for twenty-three minutes, until it was overcome. Venera 8, in 1972, provided the first temperature and pressure readings from the ground, confirming what was known earlier from radar. Venera 9, in 1975, was the first lander to send back images of the surface; it and its successor landers, Venera 10 also in 1975, and Venera 13 in 1981, sent back a variety of pictures as well as atmospheric and geochemical data. (Veneras 11 and 12 in 1978 landed successfully but failed to send back images because of problems in opening their television view port.) In 1985, two other landers, Vegas 1 and 2, sent back more information; their cameras, however, had been eliminated when the decision was made to allow the main part of those craft to fly on to Halley's comet; the trajectory to Halley's meant the Vega landers could alight only on the dark side of the planet. (The weight saved, of course, permitted more instruments to go on to the comet.) The Vegas also deployed two balloons in Venus's atmosphere. Collectively, the Soviet landers revealed a flattish landscape littered with flattish rocks; for the most part, their composition was basaltic, such as the primitive rocks which on Earth well up from the mantle beneath Earth's crust at the mid-ocean ridges to form the ocean floor.

In the exploration of most planets, landers are normally preceded by orbital mapping missions. However, because of the complexities of designing a radar mapper to be flown aboard a spacecraft, this process was reversed on Venus. The first attempt to map the planet from orbit was our Pioneer 12, known as Pioneer–Venus, which (along with its sister ship Pioneer 13, which dropped a probe into the atmosphere) was launched in 1978. Pioneer–Venus did not have the high-resolution synthetic-aperture radar, or SAR, carried aboard later spacecraft like Venera or Magellan. It had a less efficient radar mapper and altimeter, and consequently its map was composed of points of radar reflectivity which were twenty-three to seventy kilometers apart for the images and as much as a hundred kilometers apart for the altimeter, depending on the spacecraft's height above the planet. (As all Venus radar mappers have been placed in highly elliptical orbits that pass near both poles, their distance above the ground varies, and hence the range of resolution. And because Pioneer–Venus was stabilized by spinning around its long axis, unlike the later mapping missions, the radar could bounce waves off the planet less often.) The two Soviet spacecraft, Venera 15 and 16, launched in 1983, carried an SAR as well as an altimeter and other instruments; the SAR had the advantage of providing the images of objects on the surface instead of just their altitudes. Its resolution was one to two kilometers, a twentyfold improvement over Pioneer–Venus's. The two Veneras between them mapped only the northern hemisphere, down to the thirtieth parallel, as opposed to Pioneer–Venus, which covered virtually the whole planet—as would Magellan.

Magellan is the impoverished offspring of an earlier gleam in NASA's eye, a far grander mission, the Venus orbiting imaging radar (VOIR), planned in the early 1970s. VOIR would have been on the same extravagant order as the Viking missions to Mars and the Voyager missions to the outer planets, which were conceived at the same time—a period, after the Apollo years, when NASA's eyes were bigger than its pocketbook. Like Voyager and Viking, VOIR would have bristled with scientific instruments. VOIR was killed in 1982 when its cost estimates were over $800 million, but it was resurrected a year later with a slender budget of $270 million, under a new name, Venus radar mapper. Later it was rechristened Magellan. Compared to VOIR, Magellan is stripped down and economical. To save money, Magellan was patched together from bits and pieces of other spacecraft, like an old jalopy. Its big antenna was an extra one from Voyager, its thrusters also came from Voyager, and so did its octagonal equipment bus. Its computers were an extra set from Galileo, and part of its radio came from Ulysses. Its solid-rocket motor was an adaptation of a rocket used for launching satellites from the space shuttle. Even so, it overran its slender budget—largely because of the Challenger accident, which caused its schedule to be stretched by a year. Again because of the accident, the project had to go to the further expense of buying a different upper-stage rocket for the trip from Earth orbit to Venus. (The original rocket, a liquid-fueled Centaur, was judged unsafe to carry aboard a space shuttle.) The cost of its research and development and its launch, plus operations for mapping the planet once, was a little over $550 million.

Magellan has only one instrument, the SAR. Magellan would map Venus with a resolution between 120 and 300 meters, almost a tenfold improvement on the Veneras. Magellan's radar data would be transmitted in digital form through space to Earth in such quantities that scientists would receive as many bits of information about Venus as had been transmitted from all previous planetary spacecraft—from the entire fleet of Rangers, Surveyors, Lunar Orbiters, Mariners, Pioneers, Vikings, and Voyagers—put together, with all the data from the Soviet missions, the Lunas and the Luniks, the Veneras and the Vegas, thrown in for good measure. The radar had been turned on and tested in space during the cruise from Earth; it had worked perfectly—though the final proof of success, images, of course were lacking. The first radar tests of the planet were scheduled to begin on August 16, six days after VOI.

. . .

LATER in the morning after the VOI burn, in order to find out a little more about the Venus landscape, I walked on through the engineers' office space and through an area for scientific administration before crossing a hall and entering the inner sanctum of the science team itself. The team has about fifty members, composed of principal investigators and guest investigators (a distinction a little like that between full and associate professors), and many of the members, most of whom are at universities where they are professors of geology, geophysics, geochemistry, or planetary science, have little retinues of graduate students. However, very few of the students were around yet. One scientist who was present was James W. Head III of the department of geological sciences at Brown University, a trim, graying geologist who reminded me of a thinly sliced, well-dressed block of fieldstone—he wore neatly pressed blue jeans and a plaid shirt, the uniform of many of the geologists.

Head, who was vice chairman of Magellan's geology and geophysics task team as well as chairman of the team's geology and tectonic-processes group, loaded me down with maps: a small gray map which combined the best of all the previous mapping data from Arecibo, Pioneer–Venus, and Venera data. He also handed me two large, handsome polar maps of Venera data supplied by the Russians—one was in searing oranges and yellows, lest anyone forget they were looking into an inferno.

The smaller gray map, which Head unfolded for me, was easier on the eyes than the orange one, but it was somewhat fuzzy, because of the poor resolution of Arecibo and Pioneer–Venus; the resolution was markedly better in the northern third of the planet because of Venera's superior imagery. (The radar at Arecibo is constantly being perfected and now has a resolving power of one to two kilometers, approaching and perhaps even matching the Veneras' resolving powers.) The map was tantalizingly sketchy; despite the efforts of an artist with an airbrush, it had Terra Incognita written all over it—or would have had, in an earlier century. Still, Venus's equator wasn't hard to find, because it is marked, roughly speaking, by a globe-encircling chain of highlands split longitudinally here and there by central troughs. Inclined to the equator at an angle of about twenty-one degrees, the chain is broken in places. By far its biggest piece is Aphrodite, over 10,000 kilometers (6,214 miles) long, which looks like a giant scorpion covering a quarter of the planet's circumference. The scorpion's head and body were Ovda Regio and Thetis Regio; its long tail was eastern Aphrodite, a long line of grooves and ridges ending at Maat Mons, the sting. West of the scorpion's pincers, a semicircular mountain range called Hestia Rupes, is another highland area, Beta Regio, which the scorpion appeared to be menacing.

In the northern hemisphere, running from the sixtieth to the seventy-fifth parallel and filling about half the planet's circumference at those latitudes, is another huge highland area, Ishtar Terra, the size of Australia. On the west is a high plateau, Lakshmi, which rises on steep escarpments to an altitude of three kilometers. It in turn is dominated by a great mountain, Maxwell, with an altitude of eleven kilometers, towering eight kilometers above Lakshmi. (On Venus, altitudes are measured from the mean radius of the planet. If a similar point was used for Earth, it would fall some three kilometers below sea level—the figure that has to be added to terrestrial altitudes to compare them with Venusian ones. The distance between the planet's highest point, the top of Maxwell, and the lowest, the bottom of a trench called Diana Chasma three kilometers below the mean radius, is fourteen kilometers, which is less than Earth's twenty kilometers.) Beyond Maxwell is a lower plateau of broken terrain which geologists have called tessera ("tile" in Latin) because it reminds some of them of a complex mosaic floor. Tessera plateaus, unknown on other planets, are a common feature of Venus. North of the Ishtar highlands are a lot of strange grooves radiating roughly from the pole, as though they were longitude lines. Together with the belt of mountains around the equator, Venus looks as if it had been turned out by Rand McNally—with a sign of the zodiac, Scorpio, transplanted from a celestial globe.

Most of the rest of the planet is lowland plains made of the basaltic flows identified by the Venera landers. The lowlands are dotted with features—there are other mountains, plus a variety of round, dome-like mounds peculiar to Venus which geologists have come to call coronae, normally between a hundred and a thousand kilometers in diameter, which are dotted all over the planet. The planet is sprinkled with impact craters—though not as liberally as the moon or Mars, where they increase in number geometrically as they decrease in size. On Venus there is a sharp cutoff in craters below a couple of kilometers in diameter, reflecting the smallest-size asteroids or comets that can make it through Venus's atmosphere—all the smaller ones vaporize.

Skimpy as the Arecibo, Pioneer–Venus, and Venera data were, the scientists had thought a good deal about it, particularly over the last decade. Head and a younger colleague, Larry S. Crumpler, a postdoctoral research associate at Brown, had proposed a theory, in a paper they had published in *Science* in 1987, that Venus had crustal spreading—a component of plate tectonics. And they had continued this line of thought in other papers, the most recent of which had just been published in the August 9 issue of the British periodical *Nature*, to coincide with Magellan's arrival at Venus.

The field of planetary science is a close-knit one in which many of its participants have studied under each other in a sort of scientific laying-on of the hands. Head had received his doctorate in geology at Brown in 1966 under Thomas A. Mutch, then the chairman of the department of geological sciences, an avid geologist only ten years older than Head, and one of the most spirited planetary scientists of his generation; he built up a planetary group within his department and also was leader of the imaging team for Viking in 1976. Mutch had first become fascinated with planetary science in the same year that Head was finishing his doctorate, and he carried Head along with him, and some others as well. He didn't have to twist Head's arm very hard. Head, who was born in Richmond, Virginia, in 1941, was sixteen at the time of Sputnik; as a radio ham he had tuned in on its signal, having learned its frequency by writing Moscow. After graduating from Washington and Lee University, Head had gone on to Brown for graduate work in geology, with the idea of making some sort of a living out of rocks. "Tim Mutch, being tall, we always thought of him as having his head in the clouds, and we were right," Head said. "Here we were, in 1966, working on our doctorates and thinking about jobs in oil companies. In one of our last seminars before the end of the spring semester, Tim was looking out the window. 'There's no problems left in the earth's stratigraphy,' he was saying, sort of to himself. He said he saw the real frontier in geology in space. He brought Eugene Shoemaker, who was the first geologist to study the planets—he had founded the branch of astrogeology of the geologic division of the United States Geological Survey in Flagstaff, Arizona, in 1960—to talk to us about craters on the moon and on Mars. It made it kind of tough to fill out our job applications." In 1968, while he was still working on his Ph.D., Head went to work for Bellcom, Inc., which was under contract to NASA to study possible landing sites on the moon. Head returned to Brown and joined its faculty in 1973. When Mutch moved to Washington in 1978 to become Associate Administrator of NASA for Space Science, Head took over the leadership of Mutch's planetary group in Providence. Mutch was killed in 1980 while leading several Brown students on a climb in the Himalayas. Head—very much in the pied-piper tradition of his old mentor—had brought with him to

....

Pasadena a contingent of a dozen Brown graduate students.

I asked him to tell me about his work with Crumpler and the possibility of crustal spreading on Venus. In the early stages of exploring a planet, scientists often reach for what they know best—a terrestrial analogy —and, of course, the model for spreading that Head and Crumpler started with was Earth's. First, he explained, the term "crustal spreading" is a commonly used misnomer; it is not the crust but the lithosphere that spreads and inches across Earth. The lithosphere is the crust plus the uppermost layer of the mantle, which together form a rigid outer unit some seventy kilometers thick; the lithosphere in turn rides on a mushy zone of the mantle called the asthenosphere, which extends several kilometers down. Below that, the mantle becomes stronger again. (The distinction between crust and mantle is a chemical one, but the lithosphere is a physical definition, based on the similar high rigidity of the crust and the upper mantle.) The mantle constitutes about eighty-four percent of Earth's volume. In cross section, Earth, about 6,370 kilometers in radius, has a crust which varies between about thirty-five kilometers thick under portions of the continents and about ten kilometers thick under the oceans, a mantle 2,900 kilometers thick, and a core 3,490 kilometers in radius. The cross section of Venus was thought to be broadly similar, and if Venus was built the same way as Earth, and made largely of the same stuff, maybe it worked the same way, too.

On Earth, the lithosphere is broken into fifteen-

....

odd plates which are continually being formed at one side and often pulled down—or subducted—at the opposite side, where one plate sinks under a neighboring plate. Plates are formed on both sides of a giant rift which wanders—a little like the sutures on a baseball—some fifty thousand kilometers around Earth, making it the longest single feature on any of the terrestrial planets. Most of it is under the ocean, where the upwelling lava forms what are called mid-ocean ridges, with the rift running down the center. The ridges fall away symmetrically in both directions as the plates on either side separate and cool. On land it forms rift valleys. In one case—Iceland—the mid-ocean ridge is lifted above water by an upwelling in the mantle called a plume, of which more later. A series of what are called transform faults cut away from the rift in either direction at a steep angle to the rift and parallel to each other; they reflect cracks in the plate as a result of different stresses at different parts. This, Head told me, is crustal—or rather, *lithospheric*—spreading.

The fifteen-odd plates that constitute Earth's lithosphere are formed at one edge and many are subducted under the adjacent plate at the opposite edge. This was first determined because of a discovery by two British geophysicists (one of whom, Head told me, was Dan P. McKenzie, a member of the Magellan science team) that the plates are rigid, so that any movement at one side would also happen at the other—in other words, if the edge at the rift moved ten miles to the west, the far edge must, too. The rate of movement is very slow, only a few centimeters a year—but that is fast enough so that no part of the lithosphere (excluding the continents) is more than 200 million years old, juvenile compared to the surface of the moon, some of which is more than 4 billion years old. Subduction usually happens in deep trenches beneath the ocean. Mud and clay are carried from land to water by rivers and lubricate the plates, allowing one to slide beneath another—it is principally the downward pull of the sinking plate sliding into the mantle, and perhaps all the way to the earth's core, that pulls the plate across the earth's surface. Even so, the heavy plate might not budge were it not for the mushy asthenosphere, composed of nearly molten rock, which also acts as a lubricant. It is likely that some of the water subducted with the plates permeates the top of the mantle, where it lowers the temperature at which the rock becomes slushy.

On Earth, the lithosphere gets recycled. The cooler subducted basaltic material making up the crustal layer of the plates is cooked again in the mantle, driving off trapped ocean water. This water heats overlying rocks, creating quartz-rich melts of granite. Granite melts are lighter than their surroundings and hence rise to shallow depths and solidify, making the continents, which ride on top of the plates like a head of foam on a glass of beer; as continents are never subducted, they are far older than the ocean floor. On Earth, there would be no continents without plate tectonics—and there would be no plate tectonics or granites, many geologists think, without water to make the mushy asthenosphere. The subducting plate exerts a great compressive force on the upper plate so that high mountain ranges often rise on continents beyond subduction zones—such as the Rockies, the

Andes, and the Himalayas. As a result of the rising granite melts, there is apt to be an arc of volcanoes (an arc is the form a piece of Earth's spherical lithosphere takes when it is pushed down) given to highly explosive eruptions because the granite melts contain an abundance of subducted water and other volatiles which, when they near the surface, where pressure is less, turn suddenly to vapor. This is what happened at Mount St. Helens. If the arc of volcanoes happens at sea, it results in an arc of islands; island arcs eventually crash into continents, building up their mass.

How much of this scenario, if any of it, could be transported to Venus was hotly debated among the Magellan scientists—the Brown geologists versus almost everybody else. Was there crustal spreading? Were there discrete plates? Did they move across the surface, disappearing at subduction zones? Would the highlands—Ishtar or Aphrodite—turn out to be granitic, like the continents on Earth? Head believed that there was already evidence that the spreading part of the story, at least, was valid for Venus. On a big gray table in the science team's room, he unrolled the composite map based on data from Arecibo, Pioneer–Venus, and Venera. Unfortunately, he said, most of the evidence for spreading was south of the sharper Venera imagery, in the globe-encircling equatorial highlands and in particular the scorpion-like Aphrodite region, where there was only the fuzzier Arecibo and Pioneer–Venus resolution. Part of the excitement of Magellan with its superior imagery of the Venusian plains, he said as he placed ashtrays and a bowl of paper clips to hold down the curling corners of the map, was that it might afford us a view of what our own ocean basins would look like if they were drained. Head pointed to a groove that ran through much of the equatorial highlands, and in particular through eastern Aphrodite, that he speculated might be a rift such as the one running down the middle of the mid-Atlantic ridge. If so, the terrain should slope symmetrically at either side in an easily recognized profile, more steeply at first and increasing gradually until the slope vanished into the surrounding plains. The Pioneer–Venus altimeter suggested to Head that this was the case, though with measurements at twenty-three- to thirty-kilometer intervals it was too crude to be sure; Magellan's altimeter would take measurements four or five kilometers apart. Head suggested that a flat plateau on top of western Aphrodite—the upper body of the scorpion—might prove to be analogous to Iceland, where the mid-Atlantic ridge surfaces amid spectacular signs of volcanism. If so, Magellan would see them. Elsewhere in western Aphrodite, Head thought he saw in the Pioneer–Venus imagery what might be transform faults—apparent lines trending from the southeast to the northwest, possibly resulting from the uneven stress of crustal spreading, which can crack plates. If so, the Magellan imagery should clarify them. If the surface was traveling north and south from the equatorial highlands, then the terrain, as determined by the number of craters on it, should be getting older the farther it was from the equator. Crater counts made from the Venera imagery, which could resolve craters down to five kilometers across, suggested that in the northern hemisphere as far south as the thirtieth parallel, the limit of the Veneras' coverage, the terrain did increase in age with distance from the equator: the number of impact craters decreased from north to south. Furthermore, if the Venera trend was projected southward beyond the limits of the Venera imagery, the age would reach zero at the equatorial highlands. Magellan, which would resolve objects all over the planet smaller than the smallest craters on Venus were thought to be, would settle this question.

Head and Crumpler's published theory dealt only with spreading; it did not postulate plate tectonics—though Head clearly enjoyed playing with the idea. One major obstacle to plate tectonics, he said, was that he and most other scientists believed the lithosphere was too thin and too hot (because the great heat of the atmosphere keeps the surface temperature high) to have a rigidity comparable to Earth's. Nonetheless, plate tectonics couldn't be altogether ruled out. If they occurred on Venus, he said, then no part of the surface should exceed a billion years of age— the length of time it took, according to estimates based on the Venera crater counts, for a bit of crust formed at the equator to move to either pole. He pointed to the mountainous area in the north—Ishtar Terra with its two great divisions, the Lakshmi plateau with Maxwell Mons, and the area of uplifted tessera to the east. This zone of tessera, about the same size as western Aphrodite (the scorpion's head, body, and pincers), seemed to be compressed in giant creases, as if Ishtar were being squeezed like the Himalayas where the plate carrying India subducted under another plate carrying Tibet, crumpling it. And he had found a possible candidate for a subduction zone on the northern escarpment of Lakshmi Planum where the Freyja Mountains fringing the plateau plunged to the arctic plain below. There, the Pioneer–Venus altimeter seemed to suggest that the plain bulged upward in a manner characteristic of lithosphere on Earth at the point where it was about to subduct. Beyond the Freyja escarpment, mountains folded upward like, say, the Andes, as if they were being compressed. Magellan's altimeter would get much more detailed data about the bulging lithosphere. (The upward buckling, Head cautioned, could have other explanations, such as the flexure and rebounding of the lithosphere under the great weight of Ishtar.) He said he would also be looking for signs of volcanism in the Freyja Mountains—not so much for explosive eruptions, like Mount St. Helens, for Venus had no water; but for quieter signs of outpourings of lava, which might indicate a Venusian version of volcanic arc.

Another member of the science team, Roger J. Phillips, who had come in to pick up some maps, was listening in. Head introduced him as someone who was not convinced by the arguments for spreading. "Jim's hypothesis can be tested, so it is a good hypothesis," Phillips said. "If there are transform faults, or if the terrain gets older as you move away from the equator, Magellan will tell us. That makes it good science." I had the feeling that this was about as far as Phillips would go in praise of Head's theory.

. . .

ON August 16, when it was mid-afternoon in Pasadena, the spacecraft's big antenna was turned down toward the planet and the radar was turned on for the first of a series of mapping tests. Traveling over eight kilometers a second as it approached the lowest point on its orbit, the spacecraft swooped down over the north pole of Venus, starting the periapsis pass, when it zooms toward its closest approach over the equator. Thirty-seven minutes later, after it completed its first 12,000-mile swath over the south pole and was heading back out to space on the long leg of its orbit which would take it 8,427 kilometers out to space, the radar was turned off, and the spacecraft began swinging around so that its big dish antenna pointed at Earth. (The big dish, which of course doubles for radar and radio communications with Earth, is called the high-gain antenna; the word "gain" refers to the amplification of the signal.) This and most of the spacecraft's other attitude changes were carried out not by its thrusters but by spinning reaction wheels in the forward equipment module— there were four of them, one for each of the three axes and a spare, and the spacecraft could be turned in any direction by spinning them at different rates. (Venera 15 and 16, which had no wheels but relied on thrusters, ran out of fuel in three months—the end of those spacecrafts' lifetimes. The wheels would permit Magellan a far longer lifetime.) Because the wheels are weaker than the thrusters, it generally took ten minutes for the spacecraft to turn from Earth back to Venus—and vice versa. (The wheels are more accurate than the thrusters. Their only drawback is that occasionally they build up too much spin and have to be braked or de-spun, as the flight controllers called this action. As the de-spinning would cause the spacecraft to rotate in the opposite direction, the braking was done against the force of the thrusters.)

After half the data gathered during the mapping pass had been transmitted, which took about an hour, Magellan—approaching apoapsis, where its speed slowed to a little over three and a half kilometers a second—turned away from Earth to do a star calibration, or star cal, which it does once each orbit as a way of confirming the accuracy of its gyroscopes, the way every day at noon the captain of a ship will take a sighting of the sun with his sextant. The gyroscopes—those highly electronic spinning tops—remain fixed in relation to the stars, no matter which way the spacecraft turns, and hence provide what flight controllers call a platform in space. The platform can drift, just the way a top does. If the degree of drift is fairly small, it can be corrected—space engineers call this tweaking the gyros—by making an optical observation of certain guide stars. The computer—specifically, its central processing unit, or processor—measures the difference between where the gyros say the guide star is and where it actually is. Once the number of degrees of error is known, the guidance platform can be tweaked—and, that done, the spacecraft's attitude can be tweaked as well.

For the star cal, the guidance system orients one of the short axes of the spacecraft, the yaw axis, so that it is at right angles to a plane on which there is a pair of guide stars—there are twenty of them, bright enough so that they can be sensed by the spacecraft's star scanner, which sticks out at right angles from one side of the forward equipment module. (The other short axis is the pitch axis, and the long one is the roll axis.) Then the spacecraft rolls slowly around its long axis so that the scanner can find one of the guide stars. The scanner has two slits that make a V, and it records the passage of the star across each of the slits and the time that elapses between crossings. If the star is in the right place—or, rather, if the spacecraft's gyros are accurate—not only the placement of the star on the slits but also the timing between the two slits will be just right. If they are off, the gyros will be adjusted. There is a set of limits, and if the position of a star is off by more than .07 of a degree, or if the timing of the passage between the slits is off, the processor will reject the scan, on the theory that the scanner might be looking at the wrong star altogether. If the gyros are close enough and they pass the scan, the spacecraft rotates again and does a second scan on the other guide star, for confirmation. On the fifteen-month cruise from Earth, there had been a lot of failed star cals. It turned out that the spacecraft was traveling inside a blizzard of Astroquartz from the spacecraft's Astroquartz insulation blanket. Bits of white silicon fibers were constantly flaking off and were shining brightly in the sunlight as they danced like a whole galaxy of guide stars around the spacecraft; the scanners were befuddled by them. The problem had manifested itself during the cruise from Earth, and ever since, the guide stars were selected so that they were on the shady side of the spacecraft—hence, the flight engineers called them shady stars. The problem was compounded in orbit around Venus because the planet was so bright that it, too, had to be behind the spacecraft during star cals.

The star cal took twenty minutes, and when it was successfully completed, the spacecraft—past apoapsis and gaining speed as it headed back toward Venus —turned toward Earth again and transmitted the remaining half of the mapping data, which took another hour to do. The data were received at the Goldstone antenna of the Deep Space Network, about a hundred miles away from J.P.L. in the Mojave Desert. Later the computer tapes were driven by a Goldstone secretary in a Corvette to J.P.L., where the data were turned into images of craters and lava flows, of ridges and valleys, of such detail that everyone was delighted. The radar and the ground equipment passed their test with flying colors.

After transmitting the second half of the data, the spacecraft pointed its antenna down toward Venus for the second mapping pass. When that was over, the spacecraft turned back so that the big antenna faced Earth and it broadcast half the data, as it had done before (there is a certain repetitiveness to activities in orbit that makes them more easily performed by machines than by men). Near apoapsis, with half the data transmitted, it turned away to do another star cal. It was supposed to turn back to Earth immediately afterward, and at 8:35 the DSN station at Canberra began listening for the spacecraft's signal. Except when it is using its radar to map the planet, the spacecraft constantly emits what is called a carrier signal over two of its three antennas, the high-gain,

of course, and also the medium-gain (a small horn
sticking out at a rakish forty-five-degree angle from
the forward equipment module on the opposite side
from the star scanner). On both antennas, the carrier
signal was emitted over the spacecraft's two radio
bands, X and S—a band being a range of frequencies.
(On both bands, the signal comes in two parts, the
carrier and the subcarrier. The carrier is the basic
signal when there are no data being transmitted, such
as the gentle whooshing your radio makes when the
announcer stops talking. The subcarrier is what conveys the data; it can be modulated, as if your voice
was a steady tone which you shaped into words.)

Canberra should have acquired the high-gain antenna's carrier signals on both bands at 8:38, but the
station did not pick them up. An unexpected LOS is
a sign of trouble, unlike the normal LOSes that occur
regularly every time the spacecraft goes behind the
planet or turns away from Earth for a mapping pass
or a star cal. The most likely explanation is that the
spacecraft has drastically lost its orientation—that it
might even be tumbling out of control. Because the
LOS was noticed in Canberra, the flight controllers
at J.P.L. decided that the spacecraft had gone on a
walkabout. The situation sounded alarmingly similar
to the loss in March 1989 of a Soviet spacecraft that
had just arrived near Phobos, a little moon of Mars
—it had turned away to image the moonlet and was
never able to find Earth again. Ever since, I had
wondered how the process of retrieving a spacecraft
in such a predicament—perfected by the American
space program—operated successfully.

It was about ten o'clock at night. The Magellan Ace

downstairs was the first to notice the LOS. He immediately called upstairs to notify a flight engineer who was in charge of the attitude-control system. As the problem looked serious, that engineer called in other engineers, working up a ladder until he reached the project manager, Tony Spear. Most of the Magellan engineers were at home; about fifty of them came tumbling back in during the next hour. Spear lives two miles away from J.P.L., near the Rose Bowl; he was already in bed. He grumbled, found some clothes, and arrived at about eleven. Spear and his top staff—James F. Scott, the flight director, Douglas G. Griffith, the deputy flight director, Kenneth W. Ledbetter, the deputy project manager at Martin Marietta (who was normally stationed at Denver but happened to be at J.P.L.), Hamlyn, the propulsion chief, and John P. Slonski, Jr., the chief engineer—all arrived at about the same time.

Slonski is a wiry man with a long, thin face and a mind that is said to tick over as smoothly as a computer—all in all, he had the sleekness and economy of a silicon chip. That evening, he was out on his lawn with a pair of binoculars looking at a comet. (A number of J.P.L. people spend a lot of their time at night on their back lawns, stargazing.) When he heard the phone ring, he had a premonition that something was wrong. "I just got some quick information and rang off," he said. "The severity was such that I didn't want more information on the phone—I wanted to come in. I grabbed a pair of glasses—I knew it would be a long night, and I didn't want to get stuck with my contacts.

"When I got in, I tried to get as clear and accurate

a picture as possible of what had happened. I work a little like a detective coming into the scene of a crime; I take out my notebook and get the facts. Then I try to explain to the other engineers—the project manager, the flight director, or the assistant flight director—what the spacecraft has done and what I expect it to do. I have a broad knowledge of the spacecraft's systems and what the spacecraft does when it's in trouble—in particular, its fault-protection system. And for a while my chief role was to stop people from sending up commands too soon, because I didn't want anyone to do anything that would foul up the automatic safing." Even if the spacecraft is out of touch with Earth, commands can often be sent to it over the S-band channel. Because the spacecraft's low-gain antenna—the little antenna on the tripod in the middle of the big one—can receive signals from half the sky, there is always a fifty-fifty chance of getting in. The low-gain antenna does not transmit; it only receives, and it receives only the S-band. It is the spacecraft's final lifeline to Earth.

Slonski learned, from engineers at J.P.L. and also at Denver, that the cause of the LOS was a loss of what flight engineers have come to call the heartbeat—in a spacecraft, it is a steady alternation of a signal as critical as the pulse in a human body. The two computers, prime and backup, are each divided into two parts, the command and data system, or CDS, and the attitude and articulation control subsystem, or AACS. It was the CDS that had been acting up on the cruise from Earth. It is the middleman between the flight controllers on the ground and the rest of the spacecraft computer. Like many middlemen, the CDS is unable to do anything itself; rather, it conveys any changes of instructions from the ground to the AACS and reports the AACS's activities back to Earth. The AACS does the work—it controls the guidance and navigation system, runs the star calibrations, monitors the gyroscopes and corrects them, and commands the thrusters.

The AACS itself consists of three smaller units: the processor, which is the AACS's own brain for executing whatever the CDS and the ground wants the AACS to do and also for programmed activities; the memory with all its addresses and words where the processor finds the detailed instructions for all activities the spacecraft might have to execute; and the input-output data assembly, or IODA, the message center which handles communications between the processor and the systems it is controlling or monitoring. The CDS has to know at all times whether or not the AACS is in good working order, and it does this by constantly monitoring a particular address in the AACS's memory, with a mailbox number of 6200. If everything is going well, that address should change back and forth between two words every two-thirds of a second, in a manner engineers liken to a heartbeat. "Thump, thump, thump," one demonstrative manager, David Okerson, described it to me, placing his hand over his heart and whacking it several times. Okerson, who had a considerable gift for useful similes and who along with Slonski provided me with considerable information about engineering matters, is the project engineer and the liaison officer between the Magellan office and NASA headquarters, to whom he submitted daily reports—hence, he is analogous to the CDS, just as Spear and his top associates are the equivalent of the processor in the AACS. He is a lanky man always on the move but was never hard to find because his gray head could be seen bobbing above all the engineers' cubicles. Okerson, a Rhodes Scholar who graduated from Princeton in 1969 and spent ten years studying physics at Oxford, a decade ago had joined a Vancouver electronics firm designing the SAR for VOIR; in 1989, he went to work for NASA as Magellan liaison. Phobos and other Soviet planetary spacecraft, Okerson told me, do not have a heartbeat or an arrangement whereby one part of the computer (the CDS) monitors the other part (the AACS).

As with human beings, a skipped heartbeat is a symptom of trouble, in the computer's case in the AACS—though, like most symptoms, a heartbeat loss does little to identify what the underlying trouble is. Instantly, the CDS takes action, lest the spacecraft be lost irrevocably. It has a list, called the heartbeat table, of about twenty-eight progressive steps that might be taken to save the situation; if one doesn't work, the CDS progresses down the list to the next. The heartbeat table causes changes in hardware—that is, turning one component on or another off—and some of the hardware changes can cause changes in the software.

For nine seconds there was no action, a built-in delay in case the heartbeat problem went away of its own accord. Then the CDS took the first step down the heartbeat table—it swapped the prime AACS, AACS-A, with the backup, AACS-B. The two computers could be swapped in their entirety or in any of their parts, but, whatever was switched, the result was to cause a massive restart or reset of all the systems, as if the computer were turned off and on again. In computer technology, this is known as a warm reboot, because the system actually is not turned off. A restart, or reboot, Okerson told me, is the electronic equivalent of giving a balky piece of machinery a kick to make it work. Even though memory B had been rendered inaccessible after the address problem following separation of the booster six days earlier, the restart overrode those earlier instructions; it was now on line. There is no evidence that the address problem recurred, at least at this point, but if it did, it had no bearing on what happened next. The events after separation, though, did have other consequences.

The shift to memory B triggered the processor in AACS-B to enter a part of the software program known as RAM safing—something that happens whenever a memory is brought on line which has not been used for a while and which contains an encoded message that flags the fact that it is "cold." RAM stands for random-access memory; it contains a software program that the spacecraft normally operates with and that allows it to be controlled or reprogrammed by the ground—the computer is in RAM constantly, unless there is a serious problem. (The alternative to RAM is ROM, which stands for read-only memory, which contains software programs entered in emergencies; its purpose, according to Okerson, is to handcuff the spacecraft so that it does nothing to worsen its situation.) Safing is the part of the RAM program the computers enter when the spacecraft has lost contact with Earth. (There is a ROM safing program, too.) RAM safing causes the spacecraft automatically to do a star calibration so that the gyroscopes' accuracy can be checked, and then it commands the guidance and navigation system to orient the spacecraft toward Earth.

Once in RAM safing, that program has an eighteen-hour time limit to find Earth. The timer was ticking now. First the processor of AACS-B, processor B, automatically turned the spacecraft's solar panels toward the sun, guaranteeing that the batteries wouldn't run out of power—the processor used the increasing or decreasing strength of the current generated by the panels for guidance, as if the panels were a light meter; the spacecraft moves always in the direction of increased current, until it increases no more. Then the spacecraft went into what is called a coning maneuver to find a guide star—it rolls around its long axis, and it pitches a little bit, too, so that its yaw axis moves in a way that looks like a cone, allowing the star scanner to circle a broad swath of sky, in the hopes of finding one of the twenty-one guide stars bright enough to attract its attention. It was searching for the brightest of all, Sirius, but that star turned out to be behind Venus. Memory B's scratch pad, a part of the memory where navigational data—such as coordinates for the relation of the spacecraft to Earth and Venus as well as to the guide stars—are stored and can easily be updated, of course had not been updated since the alarms following separation, and consequently it was unaware that Sirius was eclipsed. Then the spacecraft coned again in search of a second star, Betelgeuse, for confirmation. Because of the antiquated navigation data, it had an imperfect sense of where to look for the shady guide stars, and the star scanners apparently settled instead on a brightly sunlit particle of Astroquartz that had broken loose from the insulation blanket. Resetting its gyroscopes, the spacecraft then aimed its medium-gain antenna—the one sticking out at a forty-five-degree angle from the forward equipment module—at the point where it thought Earth was (but where it was not), content to wait in this way for the remainder of its eighteen hours in RAM safing. It transmitted its carrier signal on the S-band. Over the medium-gain antenna, the S-band has a beam width of ten degrees. As the X-band's width was too narrow to be helpful and it used a great deal of electricity, the safing software switched it off. (The X-band is used for transmitting large volumes of scientific and engineering data, but its beam width is narrower than the S-band's, whose rate of transmission is very low but whose beam width is very wide. Beam widths of the same bands vary from antenna to antenna, and so does the transmission rate. The high-gain antenna, which the spacecraft uses in normal operations, permits the highest data-rate transmission but only the narrowest beam, and consequently its aim has to be near-perfect—within half a degree for the X-band and within two degrees for the S-band.)

On the ground, to listen for the spacecraft's S-band signal, the Deep Space Network's stations brought up its largest antennas—seventy-meter dishes, more than twice the diameter of the thirty-meter ones used during normal operations. The spacecraft's aim, of course, was far off Earth, and it would have remained in this inaccessible state, pumping S-band signals at the stars, until the eighteen hours ran out, except that, six hours and fifty-five minutes after the timer started, AACS-B experienced what Slonski and other engineers refer to as another event—a term they use when they don't really know what happened. The event took the form of a runaway program execution, or RPE, in which the processor gets into the wrong part of the memory, which in turn directs it to other wrong parts of the memory, and the computer generally runs wild, going round and round a circle of wrong addresses. There was no heartbeat loss and no further step down the heartbeat table. No one knew why this had happened, though some engineers later thought it might have had to do with the misaddressed commands that led to the alarms after separation. Nor did anyone know what the spacecraft did during its breakdown. But, as a result, the computer exited the RAM safing program and the eighteen-hour timer turned off. "If we had known that, we would have been far more upset than we were," Slonski told me. "It was devastating."

The engineers at J.P.L., who hadn't the foggiest idea what was happening, continued to wait patiently for the spacecraft's signal. The wait would have gone on for a very long time—but six hours and sixteen minutes later, a little over thirteen hours after the LOS began, the computer suffered another event, this time a programmed one. In the course of ricocheting from one wrong address in the memory to another, the computer eventually fell into a trap that had been built into the software before launch, in case of such an eventuality. A few of the 3,200 addresses in the memory had no instructions in them—and hence would never be entered, unless the computer was misbehaving. The engineers had filled these empty mailboxes with what they called trap instructions, so that if the computer ever fell into one, the instruction would command the computer to do another restart, which caused a switch back to memory A, and to enter ROM safing, the other software program used in emergencies. The computers had experienced an RPE before launch, and the engineers had been alarmed enough to set the traps in this way on the assumption that if the computer ever got into one of those empty mailboxes, it would be because the spacecraft was running amok again.

"ROM safing is about the safest place you can be," Slonski told me. In ROM, the spacecraft's memory can be read out, but it can't be written to or altered from the ground—nor can it be corrupted by glitches in the spacecraft or radiation from the sun. (In RAM, where instructions can be changed by new instructions from the ground, the memory is not what computer engineers call write-protected and is therefore subject to corruption.) ROM can, however, respond in a roundabout manner to a few simple commands from Earth, including attitude changes, or to switch radio bands, and its scratch pad for navigational information can be updated. ROM is a simpleminded backup to RAM, but the spacecraft does not enter it if RAM is functioning correctly—if it is, RAM safing will go to the end of its eighteen-hour cycle and ROM will never be entered. The only way the spacecraft can enter ROM is if the RAM part of the memory is corrupted—or if a memory problem causes the processor to fall into a trap. And memory B, which the computer had been using at the time of the second event, had problems.

ROM safing is simpler than RAM safing, requiring no star cals. In ROM safing, the spacecraft put its long axis toward the sun—as Earth was on the other side of the sun from Venus, its antenna end faced toward, not away from, the sun. The scratch pad of memory A, which was back on line, had been updated all along, so the spacecraft knew, when it was aligned toward the sun, at what angle Earth was in relation to the sun. Then it turned sideways so that its medium-gain antenna faced toward Earth and transmitted over the S-band antenna with the ten-degree beam width. To allow for a larger margin of error, it began to cone—as it had done earlier in RAM safing when it was looking for guide stars. Ideally, the circle it made should come close to Earth, so that the DSN, listening through its seventy-meter antenna (the largest at Goldstone), could easily pick up the S-band's ten-degree-width signal. The coning spacecraft was constantly listening for the response from Earth over the same medium-gain S-band—the high- and medium-gain antennas can receive and transmit simultaneously on both bands. It took two hours to cone once around; in that period, the spacecraft's signal should have been audible to the DSN for fifteen minutes.

At 11:04 a.m. on August 17—fifteen hours after the walkabout began—the giant Goldstone antenna reacquired Magellan's signal. (Had the spacecraft not entered ROM but remained in RAM safing, its signal might eventually have been reacquired—but not for two, or perhaps even three, more days.) The ground waited for three coning cycles, to get a fix on the moment the spacecraft's waxing and waning signal would be at its maximum. Then, at 7:11 p.m., the DSN zapped it with all its power, telling it to stop coning and stay put where it was. The DSN pumped out its signal from the same antenna using its highest strength, 350 kilowatts—enough so that all aircraft flying in the vicinity had to be warned, as the interplanetary pulse could interfere with their instruments. (The normal signal from the ground to the spacecraft, transmitted over the DSN's thirty-meter antennas, is only eighteen kilowatts.)

With the signal reacquired, the next step ordinarily would have been to command the spacecraft to aim its medium-gain antenna more precisely at Earth—the weakness of the signal corresponded to the amount by which the pointing was off and hence would have served as a guide to tweaking the spacecraft's attitude. However, the engineers did not do this now; they did not know what was wrong, and they didn't want to do anything that might risk another LOS.

Without changing attitude or software, they commanded the spacecraft to transmit what is called delayed engineering data from memories A and B—it was the memories' log of the faults they had observed during RAM and ROM safing and what was done about them. Transmission took a long time because the S-band broadcasts at a very slow rate, only forty bits a second. (In contrast, the X-band transmits 268,800 bits per second of radar data over one channel and simultaneously an additional 1,200 bits of engineering per second over a second channel.) With the delayed engineering data in hand, Slonski and the other flight controllers were able to piece together what the spacecraft had been doing on its walkabout—the heartbeat loss, the failed star cal in RAM safing, the RPE that ended in the trap instruction and the return to memory A and to ROM safing. Still, no one knew why the LOS had happened in the first place.

Each flight controller began looking back at his or her own area for clues. For example, the computer engineers quickly detected that the spacecraft had been using processor B as well as memory B, and they wondered whether the processor might have been responsible for the craziness that led to the RPE and ROM safing. Even though processor B had tested well before launch, it had been implicated in a problem on the ground; it had not been used on the entire trip and was therefore something of an unknown quantity. Cynthia Haynie, an engineer from Denver who was in charge of sequencing—writing, encoding, and taping the series of commands to be sent to the spacecraft—began searching through a foot-high stack of printouts of commands sent in the last week to see if a sequencing error might have been responsible for the LOS. While Haynie and the other Magellan engineers were analyzing the problem, the Galileo engineers were looking anxiously over their shoulders. Galileo, which had rounded Venus on February 10, 1990, and would round Earth in December 1990—and again in 1992—on the first leg of its complicated seven-year voyage to Jupiter, had the same computers as Magellan. While they thought and talked, the spacecraft remained pointing slightly askew toward Earth and in ROM.

Five days later, on August 21, at 7:03 in the evening, the Canberra DSN station lost its lock on the spacecraft's S-band. Three minutes later, it reacquired the signal, but its strength was dropping, which meant that the antenna had wobbled back into Earth lock and was wobbling out again; after forty-five seconds, the signal could no longer be heard. The DSN brought on line a piece of equipment called a spectral analyzer that can pick up extremely faint signals. After another two minutes, even the spectral analyzer lost all trace of the signal. Seven minutes later, it picked up a hint of a signal for about one minute and then lost it altogether. The spacecraft had started on its second walkabout.

Slonski was out in his yard once again when he heard the phone ring. He had a sinking feeling about the spacecraft, but the call turned out to be from a neighbor. He talked for a minute, hung up, and was heading back outside when the phone rang again. "I thought it was my neighbor calling back, so I wasn't prepared for the real event," he said. "I sat on the line for ten minutes, hoping the problem would go away—like a call from the DSN saying, 'Ooops! Our fault! The spacecraft didn't go away after all.' " Flight engineers are apt to remember well what they are doing when bad news comes—the way most people do in the same situation, such as where they were when President Kennedy was shot. This LOS was much more serious than the first one, because the spacecraft was still in ROM safing, the safest of safe havens, and if anything went wrong now, there was no other refuge. Just as spacecraft computers can get all fouled up, the engineers that try to unfoul them can get off on the wrong foot, and that would happen several times that night. When Slonski got in, he heard a lot of talk from other engineers that the computer had gone berserk again—another software problem, such as an RPE—but Slonski wasn't sure. "Some data I got argued against that—the way the signal fell off argued that the spacecraft was moving away and back to Earth at a controlled speed. This was not symptomatic of a berserk processor. There was a problem, all right—but at least it seemed that the computer could still control the spacecraft's attitude." If the computer was still capable of controlling the spacecraft, and if the spacecraft was still in ROM safing and was coning for Earth, then the signal should be reacquired within two hours—four at the most, if the fifteen minutes that the spacecraft's signal swept Earth occurred when the spacecraft was behind Venus. There was nothing else to do but wait. Two hours went by, and then four hours, and there was no signal.

Okerson, Slonski, and most of the other engineers, both at Denver and at J.P.L., spent that night at their posts. Engineers in the two places conferred over their squawk boxes. Slonski's belief that the computer was still in control of the spacecraft caused them to think in terms of a hardware failure—and hence to

look for a command to switch off the faulty equipment. "Finding the right switch and flicking it off
turned out to be a false trail, but it served to calm
everybody down and get them thinking about what
could have gone wrong with the hardware and what
one or two simple commands might fix," he told me
later. "We had to think in terms of trial and error,
and the process of elimination: What command might
solve A, B, and C—but not D? If that command
didn't work, we'd know the problem was D, and we'd
know how to fix it." David Okerson, the project
engineer, told me later: "The situation was like playing chess blindfolded when your opponent isn't and
knows everything that's going on. And you are trying
to plot moves to force moves when you don't even
know what pieces are on the board, and your opponent does."

The engineers held back on sending any commands
until four hours had passed—and with it the chance
of success for ROM coning, if the spacecraft was still
in the ROM-safing program. The spacecraft in fact
was not coning. Slonski and the other engineers were
following a wrong star—a bit of theoretical Astroquartz. Later the delayed engineering data revealed
that the LOS had been caused not by a hardware
failure but by another RPE and heartbeat loss, which
had caused the spacecraft to enter the second step of
the heartbeat table. This should have caused a restart
and eventual reentry into ROM safing, but it did not
because the computer was suffering what amounted
to a massive coronary thrombosis—there were several
other heartbeat losses, resulting in nine further steps
down the heartbeat-loss table.

What the engineers, in their ignorance of the true state of the spacecraft, had decided to do was to command the spacecraft to switch IODAs, the component of the AACS which communicates its instructions to other spacecraft systems. This would cause an immediate restart, the first-order, all-purpose solution to most computer problems on Earth or in space—the equivalent of electric-shock therapy in the coronary unit. The engineers in Denver and at J.P.L. disagreed about the best way of doing this—the IODAs could be switched by a primary method, which Slonski favored, or a backup method, favored by Denver. Both sides caucused. "While caucusing, I realized that something else was going on that made the Denver command right—for the wrong reasons. So we called them back and told them we agreed with them, but not for the reasons they had given. They said fine—as long as we agreed, they didn't care why."

At about two o'clock in the morning of August 22, seven hours after the LOS, the command was sent— it was as short as it could be written, and administered with the 350-kilowatt punch of the DSN's biggest antenna. There was a fifty-fifty chance that it would get in, via the low-gain antenna. "In a situation like this, you want to give it your quickest, best shot, in the hopes that it will get in," Slonski told me later. If the spacecraft was tumbling, the low-gain antenna (or any antenna, for that matter) would not be pointing at the same part of the sky for very long. The command was sent twice.

Too late, the engineers at J.P.L. remembered something—under certain circumstances, an IODA swap can also stimulate a heartbeat loss, and the pathways for this had not been disabled. "So we called Denver back and said that the IODA swap had a serious problem. Additional commands were sent up to turn off the pathways to the heartbeat loss. Even so, there is some reason to think that before the correction, the first commands got in, and that one or two of the many heartbeat losses that evening, which we found out about later, were caused by the IODA swap."

In Denver, one computer analyst whom I talked to later, Rick Kasuda, the leader of the team responsible for attitude-control software, said his group had suspected that there might have been more heartbeat losses than the ground knew about, each one resulting in the swap of the entire AACS, and since they didn't know how many swaps, they had begun to wonder whether the spacecraft might have wound up in memory B. Kasuda had been talking with some associates who worked in the systems-verification laboratory, which had computers identical to the spacecraft's and loaded with the same programs. They had discovered that memory B might have developed a software problem in its own stored version of the ROM-safing program, as a result of that memory's failure after separation of the booster rocket, now twelve days earlier. If this theory was correct, Kasuda believed the spacecraft would indeed have entered ROM safing—as the engineers had expected it to do—but would have gotten stuck in its initial phase. It could not have turned its long axis toward the sun, which it had to do preparatory to coning; instead, Kasuda imagined, it would be swinging back and forth through the sun, like a pendulum. It would be losing thruster fuel as it swung to and fro. As soon as he heard this, the deputy flight director, Douglas Griffith, who was in charge, called off all further commands to the craft until the situation clarified.

Slonski came up with an even more dire scenario. "We were not clear whether the pendulum swing was a regular back-and-forth motion, like a clock's—or whether the swings were getting bigger and bigger until the spacecraft would be going all the way around," he told me. "I tend to think in a visual way about what the spacecraft is doing—we don't have a camera out there, but still I try to visualize the spacecraft, Venus, and the sun. And what I visualized now was pretty terrifying—the craft would swing back and forth at first, and then it would start to spin up, like a centrifuge, going round and round in one direction, going faster and faster, like one of those scary swings at a fair, which goes back and forth in ever bigger arcs until it goes over the top and keeps going on and on. On a spacecraft under those circumstances, things would start flying off—the solar panels, the insulation, maybe the antenna. So that's the idea I went off home with the next morning."

At length, the engineers figured out what to do. They commanded the spacecraft to switch from memory B, if it was in it, to memory A, and stay there until it entered ROM safing properly and returned to Earth. The command was sent up several times. Goldstone reacquired the spacecraft's signal at 12:03 p.m. on August 22, exactly seventeen hours after the spacecraft had started out on its second walkabout. When Slonski came back at 12:30, he told me later, "It was like waking up from a nightmare when you're not sure it's a dream, but hoping it is, and that you'll wake up, and at last I did." (Slonski, perhaps because he knew the spacecraft so well, was capable of generating worse nightmares than anyone else. At a press conference during the first LOS, after he and the other engineers had been up all night and were bleary, he speculated that the troubles after separation of the booster might have occurred because the rocket had not been completely severed from the spacecraft and was still attached to it by a wire, jerking it out of attitude. Some of NASA's public-affairs officials, who do not encourage such downbeat thinking, later referred to this suggestion as Slonski's LOS.)

The spacecraft was commanded to stop coning at 4:12 p.m. Once again, the engineers left it in that attitude, far enough off-target so that communications had to be by the weak but broad-gauge S-band. The recorded engineering data revealed that Kasuda's fears, and Slonski's nightmare, had come close to reality. "It was very close to the brink," Slonski told me later. "It could have been pushed over the edge with a wrong command."

The engineers proceeded very cautiously. Over the next few days, they turned off the star scanners to eliminate the possibility of any more erroneous star calibrations. They deleted everything in all the addresses in memory B and filled them all up with trap instructions so that if by some mischance the computer entered memory B again, it would go directly into ROM safing. (No one wanted to risk the centrifuge nightmare again.) They made sure that processor B, which they mistrusted because it not only had been in operation during the RPE that led to ROM

safing during the first LOS but was in operation when
the second LOS occurred, could not be brought back
on line. They reenabled the heartbeat-loss response.
At J.P.L. and Denver, Slonski and others began work
on further changes to the heartbeat table, which
would be implemented later. Cynthia Haynie, the
sequencer, who had only just finished working
through the printouts of commands sent to the spacecraft prior to the first LOS, had to start all over again,
to perform the same chore for the second LOS. (The
first review had revealed no command errors, and
neither did the second.)

On Friday, August 31, the flight controllers, exhausted, were looking forward to a long Labor Day
weekend. As they had done after the first LOS, they
decided to leave the spacecraft in ROM and not shift
back to RAM, the program in which they could
resume mapping, until the following week. They had
reckoned without what is called a review board—a
panel of engineers from other NASA missions that
NASA sets up to look over the shoulders of the
engineers running a mission when it is in trouble.
The review board argued that it would be dangerous
to leave the spacecraft in ROM because they might
have another LOS, the way they did before. "We met
the next day—the Saturday of Labor Day weekend,"
Tony Spear, the project manager, told me later. "The
team was exhausted. Most of them wanted to just
relax over Labor Day, and deal with the matter later,
when they were rested. We were much distressed by
the recommendation—but we voted to go ahead. We
did it—and nothing went wrong. If we'd had a

problem, we would have responded—but we'd have been exhausted."

Reportedly, Spear had been under considerable pressure from NASA, which had had a succession of problems with the space telescope, the space shuttle, and its plans for the space station, to see to it that Magellan did not fail. Spear needed no encouragement along these lines—his connection with the mission went all the way back to the early nineteen-seventies, when he had been study manager for VOIR, and he had no intention of letting Venus escape his grasp now that success was so close. "That Labor Day, everyone was very gun-shy," Spear told me when I saw him two weeks later, in the middle of September. "We were not certain we wanted to do anything risky to the craft just yet—we did not think we could go through another LOS so soon. Now we all feel better—but we know it will happen again. We don't know the source of the problem, and so we haven't fixed it. And while we are in this situation, we can never rest easy. I and many of the others don't go to picnics or parties now. We seem to want to stay close to our homes—to work in our yards only. The people we see tend to be our office friends. Whenever we go anywhere for dinner, we never take a second glass of wine. Whenever we call each other on the phone, we always say 'No problem' before we say anything else. We all have beepers. If there ever is a problem, the beepers will give the number to call plus the code 911, to signal an emergency; that way, if we are beeped and there is no emergency, we won't be alarmed. If the phone rings and I pick it up, it gets so I'm afraid to hear the flight director's voice. And I'm getting to hate this beeper." One way that some of the flight controllers coped with the strain was by putting up cartoons about characters even more anxious than themselves (such as one about a neurotic rabbit drawn in quivery lines) and articles clipped from tabloids about far more drastic situations that made their own look like a picnic (such as one headlined "Meteor Heading for Earth, Guided by UFO Invaders").

Since the first LOS, David Okerson had been writing daily reports to NASA headquarters, copies of which he gave to me later. In a précis of the entire situation which he submitted on August 31, he wrote, using terms I had as yet no inkling of, "There is some evidence that during the first loss of signal incident the AACS suffered cycle slipping for approximately 8 seconds. This would result from a software problem causing failure to complete each 30-Hertz cycle prior to arrival of the next interrupt." Something called the watchdog timer had failed to go off. (Kasuda, I learned later, had been the one to figure all this out.) It all betokened a whole deeper level of detail and anxiety to which the engineers did not yet have a good handle.

PART II

....

Scrolling Down a Noodle

MAPPING started on Saturday, September 15, two weeks later than originally scheduled. As the planet rotated slowly under the spacecraft from west to east, the mapping proceeded to the east, orbit after orbit. Beta Regio, which was about a thousand kilometers to the west, and where the mapping would have started had it not been for the LOSes, wouldn't be photographed until the end of the first mapping cycle, the middle of the following May. A complete cycle took a Venusian sidereal day, 243 Earth days or about eight months, because that was the rate at which the planet turned beneath the spacecraft's orbit, stationary in relation to the stars.

I arrived at J.P.L. on Monday, September 17, to talk to the scientists while they were looking over the initial batch of pictures. The pictures came in long scrolls called full-resolution basic image data records, or F-BIDRs, but which the scientists called noodles— each noodle, a print about twenty feet long and about eight inches across, represented a swath over 15,000 kilometers long and twenty kilometers wide, running almost from pole to pole, each one the product of a close approach of a single orbit. Every second strip started near the north pole and every second strip ended near the south pole—there was no loss of coverage, because the strips overlapped at the poles. Each day, 7.3 such adjacent strips were produced, and several of them were fastened by magnets to a long metal panel at one side of the science room. At the moment, though, there were gaps between them. The noodles that were received at Goldstone, a three-hour drive away, arrived at J.P.L. fairly fast; the ones from Madrid, which came by air to the Los Angeles airport, came a little more slowly; and the ones from Canberra had not arrived yet at all. For most inter-planetary missions, the data had arrived at J.P.L. from the DSN stations electronically, but the volume of the Magellan data was so great that airplanes and cars were far more economical. The data arrived at the radar-imaging laboratory at J.P.L. in digital form on nine-track computer tape on rolls ten inches in diameter and an inch thick; the laboratory converted it into images. They could be seen on computer screens, but the long strips of paper provided a better overview.

About a dozen scientists, some on chairs and some with magnifying glasses, were studying the long strips on the wall. The mapping had started near the north pole and cut down through the center of Lakshmi Planum—up the Freyja escarpment, across the plateau, down its southern escarpment, and then south across a flat plane toward the south pole. From pole to equator, the angle at which the surface appeared in the image strip—what planetary scientists call the angle of incidence—changed. Though the spacecraft was looking at an angle of approximately forty-five degrees to the right, the parabolic curve of the spacecraft's trajectory and the spherical curvature of the planet were greatly out of whack (among other things, the difference meant that the spacecraft was higher over the surface at the poles than at the equator), and this affected the angle of incidence of the images. Moreover, the spacecraft's and the SAR's viewing angle varied somewhat, from pole to equator, in the interests of reducing the radar's signal-to-noise ratio—noise is static—to get the best possible imagery. The straighter down the SAR looked, the better the quality of the images—at a cost of a smaller field; this penalty could be paid near the poles, where the spacecraft's orbits were closer together than toward the equator, where the orbits were farther apart and the broader field was essential. As a result of all these variables, the angle at which the ground appeared in the image was forty-five degrees at the equator but dwindled to only nine degrees from the vertical at the poles. This situation, which made life more complex for the flight engineers and image processors but easier for the scientists, would continue for the entire first mapping cycle. In the second and later cycles the spacecraft would change its attitude more to keep the angle of incidence the same from equator to pole—though at a cost of image quality near the poles. The lives of space scientists and space engineers are full of such trade-offs. They all were hoping the spacecraft would last through as many as five or six mapping cycles, though there was funding only for the first.

The scientists were focusing their magnifying glasses on the mountainous areas north and south of Lakshmi, a welter of peaks and valleys reminiscent

of satellite views of parts of the Alps, and also at a
wealth of craters, ridges, and valleys, lava flows and
volcanoes, that had not shown up on any of the
previous images from Venus. The gaps between
noodles, due to the absence of the Canberra and
some of the Madrid data, was quite tantalizing—the
noodles, and the gaps, made it seem as if the scientists
were looking at Venus through the slats in a venetian
blind. One Goldstone noodle would show the western
rim of a huge crater, and the next Goldstone noodle,
taken three orbits later, would show the eastern rim,
with nothing in the middle except bare metal panel.
The radar-imaging laboratory had, however, isolated
a few features on single noodles and blown them up;
R. Stephen Saunders, a geologist at J.P.L. who was
the Magellan project scientist, was studying a few of
these enlargements spread out on a table. While most
of the imaging team looked as though they could
easily handle a field trip among the Venusian escarpments and lava flows, Saunders had a more laid-back,
sedentary look. In 1965, Saunders, fresh out of the
Peace Corps, where he had been doing geology in
Africa, had started in the graduate program in geology at Brown, just as Head was finishing up. He
was planning to study terrestrial stratigraphy, but
Tim Mutch, that Pied Piper of outer space, soon put
an end to that. Saunders received his doctorate in
1970, one year after Head; he went to J.P.L., where
he felt the real action was because that was where the
data were being collected. In 1970, when VOIR,
Magellan's predecessor, was being initiated, he became its study scientist; no one on the mission has

been connected with J.P.L.'s Venus missions as long as he, not even Spear. In the early-morning hours of August 16, after twenty years of waiting, he was looking at the first test strip of images from Magellan when he was told by a technician that the spacecraft was LOS. "I felt exactly like an English astronomer who in 1761 traveled to India to see the transit of Venus across the disk of the sun," he told me. "India turned out to be cloudy, so he missed the transit. On the way back, he was shipwrecked. When he came home, he found his wife had married someone else and sold all his possessions. Venus is a harsh mistress."

Now that Venus had relented and opened her cornucopia of craters, volcanoes, lava-covered plains, and mountain ranges, about all that the scientists could do, Saunders said, was point excitedly at individual features that caught their attention; they were unable yet to order what they saw into a coherent whole. He pulled out a picture of an irregular impact crater that had what looked like four smaller craters inside it—as if it were a number 4 die; there was nothing like it on Earth or anywhere else in the solar system, he said. Another that caught his attention was a small intersecting network of troughs and faults in an area called Lavinia, south of the equator, that, in the image, looked like a man with very long, rubbery legs. It reminded Saunders and some of the other scientists of a cartoon character called Gumby, an animated stick of chewing gum with rubbery arms and legs, and—perhaps to lay some sort of claim on the unfamiliar planet—soon many of the scientists and some of the engineers bought little green statuettes of Gumby with gummy arms and legs, which began appearing on desks and bookcases all over the Magellan project office.

The images looked to me like ordinary photographs, but Saunders assured me that they were anything but; because they were radar images, they required a special way of looking and thinking to understand them. In a visual image, everything depends on the angle of the viewer—the camera—to the object, but in a radar image the important element is the distance of the viewer—the radar—to the object, and hence the length of time it takes for a radar pulse to reach a particular part of the surface and bounce back to the spacecraft. There can be a confusion between a high point in the background and a low point in the foreground, which are equidistant from the spacecraft. The front slope of a hill will look shorter and steeper than the rear slope, simply because the signals from the front slope will be roughly the same distance from the radar, whereas the side sloping away from the radar (assuming it is at a shallow enough angle to see it) will be at increasing distances and the signals, which will take longer in varying amounts to return, will be more spread out. For this reason, all the hills on Venus imaged by Magellan seem to have short, steep faces and long gradual slopes behind, like bunkers on a golf course. A second difference is that in a radar image the radar waves—analogous to sunlight—emanate from the same spot as they are received, and the result is as if a viewer always had the sun behind him. Consequently, there are no shadows and little relief in the ordinary sense. And a third difference is that the eye is very sensitive to color differences which might denote chemical changes and which in black and white translate into different brightnesses, but the different brightnesses in the radar images were due either to the angle of a particular part of the surface to the spacecraft (the closer to right angles the slope is to the spacecraft, the brighter it appears), or to the texture, or the composition, or some combination of the three. What made Gumby look so much like a cartoon drawing was that he was outlined entirely in white; this turned out to be slopes of the canyon walls that happened to be facing the spacecraft. The geologists eventually got the hang of all this, and were able to understand what they were seeing at a glance, but I never did. And when they got the hang of it, they were delighted with the new tool. One newly expert scientist told me he wished there were radar images from spacecraft of all the cloudless terrestrial planets and moons, as well as the cloudy ones, because they provided information that was different from the visual data but at the same time was complementary to it.

At about eleven, the scientists were summoned to a team meeting and sat down in chairs around several large tables that had been put together in a big U. The first order of business was a minute's silence for a member of the science team, Harold M. Masursky, a patriarchal figure among planetary scientists, who had died after a lengthy illness two weeks after VOI. Planetary scientists over the last thirty years have formed a loose but cohesive group whose members keep turning up in various combinations on the science teams of one or another mission, and who meet at a number of conferences every year; many of them know each other well. Masursky was particularly well known because—with only one exception —he had been on the science team of every planetary mission since the Ranger spacecraft that impacted on the moon in the early nineteen-sixties. Saunders, standing at the head of the room by a big television monitor, said that before Masursky died, some of his colleagues from the U.S.G.S.'s astrogeology branch in Flagstaff had been able to show him the first Magellan pictures from the radar tests, and he had apparently registered seeing them—news that his fellow team members seemed glad to know; to most of them, who had never before served on a planetary mission without Masursky, it made him seem more a part of the present one. I myself, however, would have a more difficult adjustment. Masursky had had a way of blending his geological knowledge of the planets with his considerable feeling for the humanities; and during the minute's silence I hoped I would be able to negotiate the intricacies of the Magellan mission without his subtle and readily intelligible explanations.

Then Saunders began running through some of the highlights of the images from the first few Goldstone and Madrid orbits. The images, the scientists felt, more than lived up to their expectations in clarity and resolution. There was a big crater with shiny ejecta fanning out on two sides like butterfly wings— on most planets, the ejecta makes a neat circle, and no one yet had any ideas of why on Venus some craters were different. The ejecta was unusually bright, even for a fresh crater on any planet. Saunders suggested that the brightness in the radar image might reflect a compositional difference—the ejecta close to the crater would have come from deep underground, and might be a different material from the basaltic flows of the lowland plains.

"That's not what we learned in Geology I," said a sandy-haired geologist, Raymond E. Arvidson, the chairman of the McDonnell Center for Space Sciences at Washington University in St. Louis, who had studied at Brown a few years behind Saunders and had been enticed there by an ad placed on his college bulletin board by Mutch. "I learned at Brown that brightness in radar meant roughness," he told Saunders.

"You got a good education," Saunders said. "But in this case it could be something else." Brightness, of course, gave information not only on the texture of the surface but on the composition as well, and to tease the two apart, the image had to be compared with another set of data received by the SAR. Between bursts of radar pulses, there was a short interval when the antenna listened not only to the echos from the SAR and the altimeter but also to the natural radio emissions from the planet itself, and these provided information about the composition. All these activities were cycled through several hundred times a second.

Next Saunders flashed up the small irregular crater, nine by twelve kilometers, that had the four little craters evenly spaced around the center. There was considerable surprise among those scientists who hadn't seen it before. "You're telling us that those

....

four craters are from four oblique impacts, all from the same direction at the same time?" asked an incredulous scientist.

Head, who was sitting near Saunders, replied, "You have to consider that many craters would be made by projectiles that had fragmented in Venus's thick atmosphere, and that the fragments would continue on nearly the same trajectories and cause complex impact patterns." Many geologists had been speculating about what the smaller craters on Venus would look like, in view of Venus's atmosphere. They knew that smaller asteroids would be vaporized; but larger bodies, although they would make it to the ground, would be broken to a far greater extent than they are in Earth's atmosphere, and the results, they knew, would be odd. Weirdness might be a factor in any crater under thirty-five kilometers in diameter because only the bodies that made craters above that size would be totally unaffected by the atmosphere. But they expected that the butterfly shape of the ejecta on some of the larger craters, too, was somehow the result of the soupy atmosphere.

At every new picture, the geologists edged nearer to the screen. At last Arvidson, who was dissatisfied with his view, suggested that they break up into smaller groups and use the monitors in the back offices, where they could enlarge and enhance features they wanted to see more closely, and maybe try to make a mosaic of parts of adjacent strips. Before they left, a computer technician who was present warned them not to mosaic on more than one computer at a time, or else the system would crash. He advised that anyone who was about to make a mosaic

....

shout "Fore!" before he pressed the button, to prevent anyone else from doing it at the same time; and then, when the operation—which should take about a minute—was complete, to shout "Five!" which would be the all-clear. (Fortuitously, the minute of silence for Masursky had been punctuated with a loud "Fore!" and an even louder "Five," shouted by a graduate student too engrossed in a volcano or crater to bother to attend the meeting—something that Masursky had been known to do on more than one occasion.) Enthusiastic shouts of "Fore!" and "Five!" —and occasionally "Six!"—filled the air for most of the rest of the week, until the technicians were able to get the particular bug out of the system that caused the mosaicking crashes.

I SPENT most of the rest of the week in the back rooms myself. A door at the rear of the main science room led to a small warren of cubicles, devoted loosely (and with many exceptions) to one or another of the geology and geophysics team's subsidiary groups. In addition to Head's geology and tectonic-processes group, there were four others: the interior-processes group, the impact-processes group, the erosional-, depositional-, and chemical-processes group, and the global-geographic map group, chaired by Saunders himself. Collectively, these five groups, which had many subgroups and a great many overlapping members, were how the scientists deployed themselves for their reconnaissance of Venus.

Scientists taking their first look at high resolution

at the surface of a planet—I have been present for such encounters with our own moon, and with Mars, Saturn, Uranus, and Neptune and their moons—tend to go through the same stages, which are characterized first by excitement, exhaustion (and sometimes bad temper); then puzzlement over the strange new features which turn up on any planet; followed by attempts (often misguided) to understand them in terms of familiar features on Earth or on other planets they have observed; then much later comes an understanding of the planet's own unique character; next, an understanding of the planet's place in the spectrum of all the other known planets and moons; and finally a deeper understanding of the solar system itself and perhaps, as an extra dividend, a new understanding of Earth. Even though Venus had been partly mapped before, Magellan was no exception to this process, because every major increment in resolution, such as Magellan's over Venera's, amounted to another new look.

In the first few days, the excitement was heightened by a sense of confusion, abetted by a bit of competitiveness, because the scientists had to figure out what the various features were, classify them, and assign them to one or another of the groups the science team was divided into. Consequently, five or six members of different groups were apt to look at the monitors and argue good-naturedly about what a feature was and who owned it. Over the next several months, the team members, like explorers everywhere, would spread out all over the landscape, staking their groups' claims to different features, or types of features—later, within the groups, some of

....

these features would be assigned to individuals. To the scientists, Magellan was the equivalent of the Oklahoma land rush—and the starting gun had just been fired.

Invariably, the computer screens—there were five of them in the different cubicles—were manned by graduate students, more excitable and sharper-eyed than the older scientists. Singly or in groups, they moved slowly down the length of one or another of the F-BIDRs, stopping whenever anything caught their attention—a reconnaissance operation they called scrolling down a noodle; scrolling along in this fashion was to them the equivalent of coasting along an unknown shore for an Earth-bound explorer. If they found a crater they wanted to examine, they could stop and enlarge it; if they wanted to connect a ridge system with its continuation on an adjacent F-BIDR, there would be loud shouts of "Fore!" and "Five!"

I looked in at the first cubicle to the right of the door, which belonged normally to the erosional-, depositional-, and chemical-processes group but instead was filled with subsets of the impact-processes group looking for large impact craters and of the geology and tectonic-processes group looking for large volcanoes—many of the scientists were overlapping members of both groups anyway. For starters, they had to decide whether each crater was volcanic or impact, and they were not having an easy time of it. As there were no large craters, they were looking at a small, nondescript one which appeared as a bright, irregular circle on top of a black lava flow.

"Well, it's not pyroclastic, but I think it's volcanic

....

79

anyway—there might be some lava coming from a fissure at the bottom," said an Englishman with straight gray hair combed smoothly back on his head; he was John E. Guest, an expert on volcanoes from the University of London who was a member of both groups. Pyroclastic explosions, violent ones which throw huge quantities of ash and cinders to high altitudes, are apt on Earth to leave a recognizable circular blanket as the material settles on the ground, or else a streak starting at the crater and getting larger if the material has been blown downwind— and Guest said he had seen neither of these signs on Venus. Another scientist—a tall, white-haired man, Gordon Pettengill, the radar expert who had been involved with the first attempts from Earth to bounce radar off Venus and who today was the principal investigator for Magellan's radar instrument—suggested that it might be an impact crater punching through a brighter surface material to a darker basaltic material below. Guest nodded; he said it was possible that on Venus, where the surface temperature was within about 500° Centigrade of the melting point of silicate rocks like basalt, the heat of an impact could more easily melt the underlying material, causing basaltic flows to well out of the crater—in which case it would be even more difficult to make a distinction between impact and volcanism. Clearly, they were on treacherous ground.

The graduate student at the computer terminal— Mark Bulmer, a University of London student of Guest's—scrolled along to another crater, where he stopped. "I see lots of streaky white lines, with a crater at the center," he said. It was surrounded by a black disk he thought was an ejecta blanket, but one edge of it went off the screen, onto the next noodle. "It looks to me like a degraded impact crater. Fore!" After about a minute, the mosaic began to appear—one stripe after another of the picture moved down the screen, as if an invisible artist were brushing it on, in the manner of artists in old Walt Disney cartoons. The two parts of the crater were together.

Alerted by the sound of the mosaic-making, several other scientists from other cubicles began streaming in—among them a tall, curly-haired man, Gerald Schaber of the astrogeology branch of the U.S.G.S. at Flagstaff, the chairman of the impact-processes group, who (like Guest) was also a member of the tectonic-processes group. Schaber, who got his doctorate in geology at the University of Cincinnati in the early nineteen-sixties, worked on the Apollo 17 radar and the shuttle radar mappers.

Schaber has a high-pitched voice and seems a little more intense than some of his laid-back colleagues— a bit of mica, perhaps, emitting occasional sparks amid blocks of feldspar—and he asked the room at large, "Is this the big-volcano group?"

"Yes," said Guest, "but all we've got so far are small volcanoes."

The English graduate student enhanced the image.

"The black area around the crater seems to be uplifted," suggested another graduate student, Brennan Klose from Harvard. The graduate students were forever spotting things before their professors.

"Maybe it's a volcanic cone—a small shield volcano like Mount Etna," said Schaber.

Guest squinted at the picture and agreed.

"Where shall we scroll to next?" said the English student at the controls. "Up? Down? Go to a different planet?"

A few cubicles down, another group huddled around a graduate student at a computer terminal was looking at systems of ridges and valleys—sometimes they were complex networks, resembling bundles of spaghetti, and sometimes they were orderly progressions, like the ridges and valleys the Merritt Parkway crosses between Stamford and New Haven.

Scrolling on a little farther, the graduate student stopped at a curiously regular piece of terrain, where two systems of ridges and grooves apparently bisected each other, making a delicate grid pattern like graph paper. There was a long silence while everyone present tried to come up with an idea. Brennan Klose, who had followed me from the crater/volcano subgroups, remarked that possibly one set of lines in the grid was caused by the ground being compressed in one direction, and the other set by being extended in the other.

Another British scientist, Dan McKenzie from Cambridge, the geophysicist who had been one of the principal discoverers of plate tectonics on Earth, said, "The trouble with compression and extension is, how do you keep the first set so regular when you form the second set at right angles to it?"

There was a pause while everyone considered this question. McKenzie, a trim, self-contained man who nevertheless seemed capable of throwing himself into a problem with great intensity, pulled his chair a little closer to the screen and said he wondered whether one set of the lines was made by the wind scouring out troughs and the other set was deep fissures filled with lava—a series of what geologists call dikes.

There was another silence.

Sean C. Solomon, a geophysicist then at M.I.T. (now at the Carnegie Institution of Washington, where he is chairman of the Department of Terrestrial Magnetism) who worked closely with Pettengill on radar analysis, said he had no idea what they were looking at. He agreed that the regular pattern might somehow be made by the winds. "Want to hear a real cop-out?" he said. "These are not tectonic features. They're sand dunes."

No one said yes. No one said no.

THE next day, I looked in at the first cubicle, where the crater group was starting a session, trying to classify impact craters. "We're still in the reconnaissance stage," Robert Grimm, a post-doctoral research associate from Southern Methodist University in Dallas (now at Arizona State University), with a red shirt and a ponytail, was saying to the group when I walked in. "We find a crater on the noodles. We try to make a rough description, to see what it is. If there's any question about whether it's volcanic or impact, we take a vote. Right?"

"No! No! No!" John Guest said emphatically. "We don't vote! We discuss it! Voting isn't science!"

"O.K., O.K.," said Grimm. "Voting has just been shelved. The point is, we have to decide which are impact and which are volcanic."

There were enough noodles now so that they had to be assigned to individual team members to survey, and Guest began assigning them. A scientist he didn't know walked in. "Are you a member of our group, or are we just using your office?" Guest asked, and when he said both were the case, Guest put him down for a noodle.

Jerry Schaber, who with some other members of the crater group had spent the last couple of days working on criteria for describing and classifying impact craters, began handing out a stack of forms. They included spaces for estimates of a crater's size, how broad its ejecta blanket was, and whether or not it had a central peak (a small mountain at the center of many large craters caused by shock-melted rock rebounding and freezing after the impact—like a raindrop landing in water, except in that case the central blip doesn't freeze). Getting the diameter was tricky, because (given the varying height of the spacecraft and the curvature of the planet) the scale of images depended on how far a feature was from the equator. Schaber produced a pocket calculator and in a few seconds announced that a crater currently on the screen was forty-seven kilometers in diameter; he offered to teach anyone how to make the computation.

Guest took a close look at the bright blanket of ejecta around the crater—it was bright in the radar image, of course, because it was exceedingly rough and blocky. It seemed smaller than similar blankets on the moon or Mars. Guest said he believed the reason was that, when the material had been tossed up in the air by the impact, it didn't travel as far because of Venus's gravity and also because of the drag of its atmosphere. He and Schaber got into a discussion of the ejecta blankets on craters on the different planets or moons: the lower the gravity, the farther out the ejecta was tossed—on the moon, rays of ejecta can circle the globe completely. Venus had the highest gravity of any terrestrial planet except Earth—and, of course, the thickest atmosphere, and therefore, the two agreed, it would be expected to have the narrowest, and the deepest, blankets.

Guest put his face in the screen to get a better look at something else he had noticed, and announced that the ejecta blanket appeared to be divided—a thick layer of blocky debris close in, and a thinner blanket of lighter material that appeared to have moved outward in lobate flows—sort of rounded peninsulas that jutted out from the blanket into the surrounding terrain. "Should we measure just to the edge of the thick, hummocky stuff, which is fairly circular, or go all the way out to the lobes on the perimeter, which are fairly irregular?" Guest asked Schaber. "However we do it, we should all do it the same way."

After some discussion, they agreed to use the hummocky inner ring of ejecta for the measurement—though they would keep track of the outer lobes, too. "The hummocky stuff is what we used on Mars and the moon," Schaber said. Planetary scientists like nothing better than comparing features from one end of the solar system to the other, something (like Earth analogies) that they resort to with particular frequency when they are taking a first look, for the sake of familiarity in unfamiliar terrain. They

went on doing it. Guest pointed out a crater with
flows surrounding it that extended outward in long
lobes. "On Mars, the ejecta sometimes formed lobate
flows, which we thought were due to underground
ice that had melted, but here we have lobate flows,
too, and there is no underground ice or water on the
entire planet," Guest said. "That opens the question
of whether the lobate flows on Mars really were caused
by underground ice. When I was a member of the
science team of the Mariner 10 mission to Mercury,
I and another scientist predicted that the craters on
Mercury would be the same size, with the same
diameter of ejecta blankets, as the ones on Mars,
because Mercury and Mars had very similar gravity.
But we were wrong, and the reason was that Mars
had an atmosphere and Mercury didn't. Sometimes
it's exciting to be wrong. We haven't thought much
about the effects of atmospheres on cratering, and
on Venus we have the most extreme atmosphere of
all."

(One person who had thought about this question
was Peter Schultz, an associate professor at Brown,
and Magellan guest investigator, whom I met later.
In his laboratory, Schultz has fired bullets into various
target materials inside chambers containing different
atmospheres and pressures, and photographed the
result with high-speed cameras. Among other things,
he had predicted the lobate flows, and he later found
the explanation for the strange butterfly patterns
surrounding craters that had puzzled the scientists
earlier: as the ejecta shoots upward after the impact,
high winds following the asteroid through the atmosphere will slam into the rising cloud of ejecta and

smash it into the ground, making the odd patterns.)

I went next door, where a group of scientists studying plains had found a long, sinuous lava flow like a river that forked downstream and was an exact duplicate of the lifeline in the palm of the hand of one of the scientists; and when I returned to the crater group, they were attempting to describe on Schaber's form the crater that looked like a 4-die— the outstanding example of what a thick atmosphere could do to an impacting body of a certain size. The four separate impacts had created one cavity that was shaped a little like a kidney.

"I think we should describe its shape as irregular," Schaber was saying, starting to write.

Guest said he preferred to call it non-circular. "We have irregularities all over the place, which are really irregular, but this one is partially round," he said.

Schaber went along with this. "Now is the time to object to terms we don't like," he said. "We don't want to get stuck with terms we don't want."

They agreed to describe the crater walls as blocky and terraced, and Guest went along with Schaber's suggestion of the word "irregular" to describe the crater floor. "It's about as irregular as they come," he said.

"Does the crater have a central peak?" Schaber asked the room at large. "Yes? No?" Someone said it had. Someone else said it hadn't.

Just then Grimm, who had gone out for coffee, came back. "Are you still working on that stupid crater?" he said. The group decided they had had enough.

.   .   .

AFTER they left, two scientists whose office happened to be in that particular cubicle, Ray Arvidson, chairman of the erosional-, depositional-, and chemical-processes group, and John A. Wood of Harvard and the Smithsonian Astrophysical Observatory in Cambridge, who was a member of the same group, reclaimed their desks. They were concerned with dust on the surface, how it was formed, and how it gets moved about by the wind or other processes. Jeffrey J. Plaut, a graduate student of Arvidson's at Washington University (who the following September became a post-doctoral research associate at J.P.L.), slid into the seat in front of the monitor and scrolled down a noodle until he found a volcano spilling either lava or pyroclastics down its cone onto the plain beyond. If it was lava, it would be of interest to the volcano and the plains groups; if it was pyroclastic material, it would also be of interest to Arvidson's group, because it would tell something about the winds.

"I think it's full of clinkers," Schaber, who had stayed behind, said.

Plaut scrolled along the noodle until he came to the picture of the alleged sand dunes. Barry E. Parsons, a onetime student of McKenzie's who was now a geophysicist at Oxford and who had been present for the original discussion about the dunes, told Plaut to stop his scrolling and asked the assembled dust people what they thought of the possibility of dunes. The alleged dunes extended from one side of the twenty-kilometer-wide noodle to the other.

"Do dunes come that long?" Wood asked.

"I was thinking more of ripple marks, like there are in the sand on the basaltic ocean bottoms," Parsons said. "They go on forever." Parsons was an expert on the oceanic abyssal plains—a good analogy to Venus, some scientists thought, because the density and pressure of the atmosphere at the surface was so great.

Someone in the room asked where the sand, or the dust, would have come from in the first place—on the moon, the dust is caused by smaller asteroids and micrometeorites, which would be filtered out on Venus; on Earth it is caused by wind erosion or water erosion, which would not happen on Venus. Venus is not a watery place, and it is not normally a windy place. (Arvidson had told me that since only about twenty percent of the sun's light, which is what drives the winds on Earth, penetrates Venus's atmosphere and since Venus's thick, soupy atmosphere takes a good deal more driving than Earth's, it lies on the ground hot and heavy and is almost as turgid as the water at the bottom of the ocean, which seldom moves—unless it is stirred by volcanism or a landslide. Big winds on the surface of Venus are extremely uncommon, and when they occur they are the result of rare volcanic eruptions and impacts.)

In an attempt to explain any sand or dust, Wood said, he had made some calculations over the summer and believed that the minerals in basalt would be destroyed in time by the chemicals and the great heat in the atmosphere, in a Venusian version of weathering. Basalt, he said, contains feldspar, which reacts with the sulphur in the atmosphere to cause a calcium sulphate, which in turn tends to swell and fracture the rock, weakening it; in time, the rock sloughs off into sand.

Another scientist remarked that ripple marks were apt to be wavy, and the lines on Venus were straight. "Not if there is very little sand," Parsons replied. "On the ocean bottoms, where the sand is thin, it forms long, straight lines."

"It's awfully regular—quite beautiful, in fact," Wood remarked. "I can't think of any other explanation."

Plaut, who believed in a tectonic explanation for the long lines and who had been looking increasingly unhappy at all this talk about sand dunes by geology professors from eminent universities on both sides of the Atlantic who he felt should know better, had left his post at the television monitor and was telephoning Brennan Klose from a phone at an empty desk at the back of the room. "Hurry on over here!" he said in a loud voice. "We need you! We're trying to get the sand dunes out of some people's heads!"

There was an uncomfortable silence in the cubicle. At last Wood said, "It's wonderful to be a graduate student!"

Klose, who was Wood's own graduate student, slid into the seat at the controls recently vacated by the outspoken Plaut. "Gentlemen, shall we look for the borders of the so-called dune field?" he asked. This required making a mosaic with the noodle next door. A "Fore!" and a "Five!" later, Klose pointed to the margin of the field, where the long dunes and troughs dissipated into lots of little ridges and valleys. "From which I deduce that the dunes are ridges and valleys, too," he said.

. . . .

. . .

OUT in the main room, the first set of F-BIDRs from Canberra had just arrived and been tacked up on the wall, filling in the gaps between the sets from Goldstone and Madrid. Several scientists were standing on chairs, among them Jerry Schaber, who had outfitted himself with a magnifying glass. He offered to give me a guided tour: "Lava flows. Tesserae. Impact craters . . . This is the interior of Lakshmi— see how high the hills are, and the slopes down the side. Here is the Planitia Snegurochka. No craters there—except maybe these circular arcs are parts of a giant ring . . . Nice volcanic feature here. A couple of domes. Here are some fractured plains. Here, this might be ejecta from a crater. Here are a pair of craters, made by an asteroid that broke up close to the ground—it was such a small asteroid that it almost didn't make it to the ground at all . . . Here's a crater with a black floor and another with a white floor. The black floor means that lava flowed into it, maybe just after the impact . . ."

Among the gaps that had been filled in were the remainders of three or four very large craters that had been first detected by the antenna at Arecibo. Donald Campbell, a short, wiry radar physicist from Cornell who had done much of the work at Arecibo, was standing a few feet from Schaber, looking at the latest crop of Magellan pictures. He said, "It's nice to have the Arecibo data as a context for the Magellan strips, even if it is at lower resolution. Jerry Schaber finds those big craters—and it's nice for him to know that they're part of a field of large craters, which I named the crater farm." The Arecibo physicists, who

. . . .

had been bouncing radar off Venus for thirty-five years, were a little patronizing toward their team-mates.

Schaber, without looking up from the F-BIDRs, told Campbell that they would all be much more excited about the crater farm now if he hadn't spoiled the surprise with his Arecibo data.

Campbell told Schaber there would be plenty of surprises later when Magellan began mapping the other side of Venus, where there was no Arecibo data; he told Schaber he bet he would miss the Arecibo data then.

The three craters in the farm were named Howe (after Julia Ward Howe, the suffragette), Danilova (after the ballerina), and Aglaonice (after a Greek priestess known for her ability to predict eclipses). The subcommittee on nomenclature for the inner solar system of the International Astronomical Union (which might have picked a more felicitous title for itself) decided that all features on Venus would be named for females. Different types of features are being named for different categories of women. The largest features on Venus are named for mythological women—regions, for example, are named for giant-esses and titanesses, coronae are named for fertility goddesses, chasms for goddesses of the hunt, and the highland terrae for goddesses of love. Smaller fea-tures such as impact craters and calderas, however, are named for ordinary mortal women—usually des-ignated by their last names. For impact craters smaller than about fifteen kilometers in diameter, first names of women from all over the world are used, such as Barbara or Ludmilla or Fatima. (One astronomer in

Hawaii contributed a list of six hundred first names gleaned from most of the cultures in and around the Pacific Basin.) Sometimes smaller craters come in pairs; and these are often given a double first name, most common in Western cultures, such as Marie-Antoinette. Craters that are more than twenty kilometers across are named for well-known women who have been dead for at least three years. About four hundred names have been nominated and put in a name bank to be drawn on as new features turn up; not all names had yet been approved by the committee (and some in fact would not be). However, so many new craters, volcanoes, and lava flows were turning up that, whether approved or not, names were needed in a hurry, so that the scientists could identify the new features in their discussions and in papers they were writing; without the names, scientific discourse would grind to a halt. According to Mikhail Marov, the subcommittee's chairman and head of the Department of Planetary Physics and Aeronomy at the M. V. Keldysh Institute of Applied Mathematics in Moscow, the name bank was in danger of going bankrupt, even with the infusion from the Pacific Basin, and he estimated his group would need another four thousand names before they are through. That is the same number as are on all the planets (excluding our own) and moons from Mercury to Pluto. The Magellan project office requested the public to submit names, and they have been pouring in—names of mothers, wives, and girlfriends; also names of cartoon characters such as Betty Boop and Barbie. So far, none of them has made it to Venus. One that might, though, came from a woman offering the name of her grandmother, who during the Depression ran a bakery that regularly fed the unemployed.

The subcommittee, Marov told me, tries to pick names that are not controversial—a term it applies with a broad and sometimes quixotic brush, for it does not wish to get into trouble. The brush is also not always predictable. Juliet Gordon Low, the founder of the Girl Scouts of America, was turned down because she was considered too political. On the other hand, Margaret Sanger, the founder of Planned Parenthood, is safely enshrined (or encratered) on Venus, unscathed by a single pro-lifer's sling or arrow. Julia Howe of the crater farm was later unfrocked; at a meeting of the nomenclature subcommittee in Florence in 1991, she was adjudged contentious and stripped of her crater, which was reassigned to Saskia, the exemplary wife of Rembrandt. (Saskia, though a first name which, according to the rules, should have been used for a small crater, was considered sufficiently identifying of the individual to be given to a large crater.) Religious women are not used, either —a Christian saint might not go over in a Muslim culture. Evil women such as Lucretia Borgia are eschewed, too, but anything else goes. Even Mary Queen of Scots got a crater, named Stuart, despite several letters from British writers asserting that she was not a nice woman and had been plotting her cousin Elizabeth's murder when she was quite rightly beheaded. (Oddly enough, Elizabeth did not make it to Venus. The input from the British was that her last name, Tudor, was not the name she was generally known by; and her first name, of course, was not

sufficiently identifying.) One member of the committee told me he would have no objection to burlesque
queens and strippers, such as Gypsy Rose Lee; he
reasoned that courtesans from the past were on the
list, so why not some recent ones? The committee,
which prides itself on its open-mindedness and diplomacy, has nonetheless landed itself in hot water more
than once. Some women's groups have objected to
what appears to them to be the segregation of women
in a sort of seraglio in the sky, and not long ago there
was an article in a San Diego paper equating the
planetary scientists' naming features on Venus for
women with meteorologists naming hurricanes for
them; the writer argued that Venus was even more
unpleasant than a hurricane, filled as it was with
steamy calderas and torrid temperatures. When I
asked Saunders about this, he argued that Venus in
many cultures was associated with women; moreover,
he found it a lovely place—in his opinion, Venus's
craters, with their pronounced white blankets of
ejecta, were among the most beautiful features in the
solar system. However that may be, I learned later
that among the nine members of the committee
selecting names for the women's planet, there was
not a single woman—the committee, in fact, was
known as the nine wise men. (Nor, for that matter,
was there a single woman among the fifty-odd principal investigators or guest investigators of the Magellan science team.)

Ray Arvidson, who was interested in Danilova,
Aglaonice, and Howe (or, rather, Saskia) precisely
because of their bright ejecta blankets, told me that
when he had first seen them—they range in size from

twenty-three miles in diameter to thirty-six miles—
he had thought they were volcanic; but quite clearly,
he and Schaber now agreed, they were impact craters.
Because of their clean-cut outlines and highly reflective ejecta, which were a sign of rough, uneroded
material, they were relatively young, he said—under
a hundred million years. This estimate—like most
age estimates of planetary surfaces—came from crater counts, which have become a universal tool for
dating all over the solar system. Obviously, the more
craters on a terrain, the older it is, but there are
refinements to that principle. Planetary scientists can
tell with some confidence how old a particular terrain
is by counting the number of craters of different sizes
and then extrapolating the age from the dwindling
number of asteroids or comets that would have hit
the planet—a curve known as the meteoroid flux,
which steepens precipitately as it reaches the period
of heavy bombardment in the first half billion years
of the solar system, tapers off less dramatically in the
next billion years, and has remained fairly constant
for the last three billion years. All planetary bodies,
of course, are ultimately the same age as the solar
system. What the scientists are trying to find out with
their dating techniques is how long ago a particular
body or a particular area of its surface was altered
into its present state, usually as a result of tectonics
or of resurfacing by volcanism—in particular by lava
flows. The flux, which is pretty much the same for
the earth, the moon, and Venus (and can be corrected
for other parts of the solar system), provides a mod-
erately accurate tool for dating a surface: the number
of large craters—large craters are fewer in number

and easier to count than smaller ones—is divided by the area being observed, yielding a relative age. The crater counts from any body in the solar system can be compared with areas on the moon from which American and Soviet scientists have received samples that are dated—the moon is the constant yardstick against which all other planetary bodies are measured, though in some cases adjustments have to be made.

The more internally active a planet is, the younger its various terrains are apt to be, because they get resurfaced more often. The oldest surface areas of the moon are in the neighborhood of four billion years, whereas parts of the surface of Jupiter's fiery moon Io are being born now—as indicated not only by the current eruptions but by the total absence of craters. The average age of the surface of our moon is between three and a half and four billion years; of Mars, between about two and a half and three billion years; and of Earth, between 350 million and 450 million years. For these bodies, the figures for the average surface age are somewhat meaningless, because different areas have widely different histories and ages. On the earth, for instance, the oldest rock of the continents have an average age of about a billion years, but the ocean basins, which make up seventy percent of the planet, have an average age of only 50 million to 60 million years, reflecting the efficiency of plate tectonics as a resurfacing agent. But the ages of different parts of Venus's surface are apparently more uniform, and consequently the average age is significant. The most recent estimate was 400 million years. Arvidson, who had been looking at the Arecibo data, told me he had arrived at the

100-million-year age for these young craters because the fraction of craters that size that had bright young halos of ejecta was a fourth, and when he divided 400 million by four, that was what he got. A hundred million years, he figured, was the length of time it would take for the jagged ejecta to smooth over so it wouldn't appear bright on the radar. (A hundred million years ago, dinosaurs roamed the earth—and would do so for another 40 million years, until they were wiped out allegedly by an asteroid that made a crater substantially larger than any in the crater farm.)

The average age of the surface of Venus has been in dispute for several years. In 1986 and 1987, two Soviet scientists, Alexander Basilevsky and Boris Ivanov on the Venera 15 and 16 team, had applied the lunar flux to the larger craters in the Venera maps and estimated the average age of Venus's surface at one billion years, give or take half a billion years. (On Venus, the craters used for dating have to be over thirty-five kilometers. Craters smaller than that, of course, are burned up in the atmosphere—the smaller the size, the greater the loss and the more room for error.) Schaber and Eugene Shoemaker, the founder of the U.S.G.S.'s astrogeology branch, thought Basilevsky and Ivanov were wrong. Shoemaker had noted that there are a greater number of craters under half a billion years on Earth than could be accounted for by the flux on the moon—where craters that age or less are harder to date precisely. The increase in craters on the earth, he felt, was the result of an increase in comets, whose incidence is a little skittish over astrogeological time; and when in 1987 he and Schaber revised the flux figures for Venus in light of the comets, they predicted a most probable average age of 400 million years (the number Arvidson had divided by four), with the oldest areas being possibly a billion years old.

Over the past twenty-five years, much of the work on the flux for the moon and all over the solar system has been done by Shoemaker. In recent years, Shoemaker, a geologist, has become something of an astronomer: he uses telescopes to study the asteroids or comets that cause most of the impacts on the inner planets. They originate in collisions in the asteroid belt between Mars and Jupiter, and as a result they can get kicked into trajectories that take them inward, so that they cross the orbits of Mars, Earth, or even Venus. Any asteroid or comet that crosses a planet's orbit can potentially impact on it. The ones that approach or cross the orbit of Earth and the moon are called, respectively, Amors and Apollos. When I saw Shoemaker at a planetary-science conference in Houston the following March, he told me that over the years he had been photographing the sky around Venus's orbit, searching for what he calls Venamors and Venapolls, his name for Venus approachers and crossers. He has found that sixty percent of Earth crossers are also Venus crossers. The dimmest he can spot is 7.7 magnitude, which at the distance of Venus would be an object about nine hundred meters across, capable of blasting a crater nine kilometers wide. (A rough rule of thumb is that an asteroid will excavate a crater ten times its own size.) At 7.7 magnitude and brighter, there are about six hundred Venus crossers; as they increase predictably in number as they decrease in size, he can compute the total number, including the smaller ones. He has figured that a given Venapoll has three chances of hitting Venus in a billion years. The cratering rate he extrapolates from the asteroids is roughly consistent with the cratering rate he deduces by applying the flux yardstick to the craters on Venus—which indeed is as it should be.

IN another cubicle, several scientists—Dan McKenzie, Barry Parsons, Sean Solomon, Jim Head, and Roger Phillips—were arguing about what was holding up the huge Ishtar area, which includes not only the Lakshmi plateau but also Maxwell Mons, at eleven kilometers the highest peak on the planet. Maxwell is named not for Elsa but for James Clerk, the nineteenth-century Scottish physicist who discovered the electromagnetic spectrum. "He is the only man in the harem," Marov told me. I asked him how this had come about. When the first crude radar maps of Venus made from Earth were done in the mid-sixties, a very few large features, such as Alpha and Beta Regiones, were visible, and these were named for letters in the Greek alphabet. Later, the nomenclature subcommittee began using names of famous scientists, and Maxwell, Faraday, Gauss, and Goertz appeared on the Venusian map. But then, in the middle seventies, radar mapping from Earth greatly improved, and with the arrival of Pioneer–Venus at the planet, there was such an avalanche of new features that Harold Masursky, who had been head of the subcommittee at the time, suggested using women's

· ·

names, of which there was clearly a suitable supply. Faraday, Gauss, and Goertz lost their mountains, but Maxwell retained his by virtue of the fact that his name had already been used in a published paper; the others had not.

Already impressive statistics about Maxwell and Ishtar as a whole were coming in from Magellan's altimeter. The steep western flank of Maxwell rises at a twenty- to thirty-degree pitch above Lakshmi Planum, a vast, flat, roughly circular plateau—in an unbroken scree extending upward for ten kilometers. For its distance, it is a greater rise than any on Earth except for some in the Himalayas. (Because of the heat of the rocks, which make them weaker, slopes on Venus usually tend to be less steep than on Earth; they are more likely to be reduced by crustal flow. The fact that Maxwell's flank is so steep led many geologists to believe it is quite young.) Mountain ranges such as Vesta Rupes and Danu Montes rose four kilometers above the southern edge of Lakshmi Planum, hemming it in with an escarpment that was eight hundred kilometers broad. The drop to the plateau floor occurred between two altimeter readings, taken five kilometers apart, so the pitch was not determined, but it was expected to be the same pitch as Maxwell's western flank. Sacajawea Patera, a caldera on top of the plateau, had a central cavity a hundred kilometers in diameter whose bottom, from Venera altimeter data, proved to be at least half a kilometer deep. (Sacajawea, the Indian squaw who guided the Lewis and Clark expedition, almost didn't make it to Venus; Indian groups thought she was a traitor to her people.) A caldera is a volcano with a

· ·

slump feature at the top, the result of a magma chamber which has drained and later caved in. A few days later, when the Magellan altimeter got a good reading of the floor of Colette, an even larger caldera near Sacajawea on top of Lakshmi, its depth turned out to be two kilometers, reaching down almost to the same level as the plains that surround the Lakshmi plateau. These low-lying plains are covered with lava, some of which might have drained from the magma chamber under the caldera. Lakshmi itself, rimmed by high mountains as if it were a gigantic frying pan, was flooded with lava from the calderas, so that its surface appears smooth; very likely, the plateau's terrain beneath the lava is a rough-and-tumble tessera, as is the rest of Ishtar.

McKenzie, Parsons, Solomon, Phillips, and Head were particularly interested in the underpinnings of Ishtar and especially of Maxwell, because they are an enormous mass to be supported on the crust of any planet. Large surface features tend to sink down and spread laterally over geological time. Venus's lithosphere, the crust plus the rigid topmost zone of the mantle, which together would support topographical features, Solomon pointed out, was thought to be thinner than Earth's lithosphere—between ten and thirty kilometers for Venus's and up to seventy for ours—and yet there were Lakshmi and Maxwell, almost as high and massive as the Tibetan plateau, the largest similar feature on Earth. The Tibetan plateau is raised up by the plate carrying India crashing into an Asian plate, but no one, not even Head, was proposing plate tectonics as the force that was thrusting up Maxwell Mons.

Head was the lone geologist in the room—the others were all geophysicists—and he may have felt a little constrained. Although geophysicists had contributed heavily to the theories of spreading and plate tectonics on Earth, many of them had other ideas about Venus. Geologists, broadly speaking, deal with the surface characteristics of planets, whereas geophysicists, who have the tools of physics at their disposal, which enables them to get at internal processes, tend to deal with what goes on inside planets. Although there is considerable overlap between them, there are differences; geologists often regard geophysicists as too theoretical in their thinking, whereas geophysicists tend to think of geologists as too observational. If the observations don't fit their current theories, the geologists think, the geophysicists would prefer to ignore the observations. Conversely, geophysicists are frequently frustrated by the fact that if the geologists' facts don't fit the theories, the geologists ignore the theories. Since the surface and interior of planets are closely linked, the two types of scientists have to work together, and the results are not always harmonious; scientists tend to think differently and favor different types of evidence. Head tries to bridge the gap and enter into the geophysicists' way of thinking, more than most geologists.

Head asked whether a glob of rising magma in the mantle—which geophysicists call a plume—might be holding it up. McKenzie agreed. "I don't see how you are going to get Maxwell up there without some sort of a column to support it," he said. A plume—hot material rising through the mantle—would be such a column.

The discussion became quite technical geophysically, and later I got a gloss on it from Solomon and also from Roger Phillips, the head of the interior-processes group. "There is something called isostasy, a concept which has been around for more than a hundred years," Phillips told me. Phillips is a geophysicist who was then at Southern Methodist University in Dallas but is now at the McDonnell Center for Space Sciences of Washington University in St. Louis. If Head, with whom Phillips often disagreed, had reminded me of a thin slab of fieldstone, Phillips was more like a piece of basalt, a little chunkier and with blacker hair than Head, but nonetheless with some sharp edges. Phillips is an ex-football player for the Colorado School of Mines; after getting his doctorate in applied geophysics at the University of California in Berkeley in the late nineteen-sixties, he was team leader for the Apollo 17 radar sensor that Jerry Schaber had worked on, and later was involved with the Pioneer–Venus mission.

"The theory of isostasy suggests that on Earth, and on other planets too, variations in surface topography will be balanced by variations of rock density in the interior," he told me. Phillips teaches undergraduates as well as graduates, and he was inclined to lecture me as if I were a freshman—something I needed when it came to the murky depths of planetary interiors. "In particular, no vertical section, from upper mantle to crust, will be more massive than any other—any excess mass in one place will flow away, and everything will be in equilibrium. A corollary of the theory is that any big topographic feature like a mountain has to be compensated for, and held up by, what we call a mass deficiency thirty or forty or fifty kilometers beneath the surface which balances the mass excess of the mountain. Like a bubble rising in the water, it is pushing up on the mountain, which is heavier and pushing down; also, it balances the mass excess of the mountain, so that the two together are in equilibrium. Nature likes to compensate in that way because it likes to minimize stresses—as do we all! What we call the depth of compensation is the point at which the mass excess and the mass deficiency balance each other out. This point can be identified from gravity readings, which are obtained by carefully tracking the spacecraft's signal to ascertain small variations in its speed, which in turn represent the varying gravity below.

"With the classic theory of isostasy, there has always been a problem when it comes to supporting the biggest surface structures, because they often have large depths of compensation. The deeper you put the depth of compensation, the hotter the planet gets. If you put it too deep, the lower-density compensating mass will simply flow away horizontally. The depths of compensation for some of the huge features on Venus like Beta Regio are so deep that mass deficiencies in the old sense just cannot be sitting there holding up the mountain. Therefore, we think that plumes, which are hotter and less dense than the surrounding mantle, must be doing the compensating. You can imagine an upward-flowing plume, which because of its low density is itself a mass deficiency, holding up a broad topographic rise like Beta. Incidentally, if that is the case, it is an indication that Venus has no asthenosphere, because if there is one, and if a plume is supplying the mass deficiency that is holding up Beta, the mushiness of an asthenosphere—which on Earth allows the plates to slide across the mantle—would cause the plume's upward thrust to be strongly muted."

This brought us to the subject of the plumes themselves.

THE theory of plumes was first proposed by W. Jason Morgan of Princeton in 1971—it is a little more than half a decade younger than the major proofs of seafloor spreading and plate tectonics, which were discovered between 1961 and 1968. Plumes are vast globs of hot rock that rise through the mantle—an odd concept to me, in view of the mantle's solidity. Indeed, I was still having trouble with the notion of subducted plates sinking through the mantle perhaps to the core. "Let me lead you through a series of steps," Phillips said in his professorial manner, which I had come to value. "Water flows easily. Syrup is a little more viscous—sticky—and it flows more slowly. Ice in glaciers flows, but it's basically solid. What you have to come to is a realization that the mantle has an even higher viscosity, but it will flow on a very long time scale. It's a rock—you pick up a piece of it and drop it on your head, you'll know it. And it transmits seismic waves, so it is solid from that point of view. But if you kind of squeezed on it slowly for a very long time, you'd have to accept the notion that, while the mantle is solid on a short time scale, on a very long one it flows, behaving like a sluggish fluid."

I asked Phillips to tell me about the plumes. "Ever see a cloud coming from a volcano?" he asked. "That's a plume. It's something of lower density than its surroundings, rising through it. Inside the earth, think of the plume and the mantle as two different fluids. Now think of the bottom of the mantle, at the boundary with the core—it is what we call a thermal boundary layer. The core, which is composed of an iron alloy, is very hot—the solid inner part of the core is about 6,000° to 7,000° Celsius, and the outer, liquid part ranges outward down to 4,000° or 5,000° Celsius at the core-mantle boundary. The core loses heat by transferring it to the mantle. Consequently, the bottom of the mantle, in the thermal boundary layer, is hotter than any other part of the mantle, which gets cooler as you go up. Because the lower mantle is hotter, it's less dense than the mantle above it; the least dense part is in the thermal boundary layer. When you put a fluid of lower density below a fluid of higher density, it is inherently unstable. The lower-density fluid wants to break out through the higher-density fluid. And it will break out. There are a lot of examples of plumes in nature, like a pot of water on a stove. When you first turn on the flame, the water just sits there and the heat passes through conductively. But as the water gets hotter, it gets unstable, and to get rid of the excessive heat at the bottom of the pot, the water there actually has to rise."

At their source, the plumes start by breaking out of the thermal boundary layer as very narrow columns, but as they rise in the mantle they broaden. Sometimes the tops break off from the stems and go on rising on their own, like bubbles; Saunders calls the phenomenon blob tectonics. Plumes rise through the mantle at a speed of ten to twenty centimeters a year—somewhat faster than the plates move tectonically across the earth or are subducted through the mantle. The plumes cover the 1,800-mile distance from core to the base of the lithosphere in about 20 million years. When they hit the bottom of the lithosphere, they are apt to flatten out against it— resembling mushrooms. If they are big enough, they bulge the lithosphere upward into a dome. One way of measuring their force and the amount of heat they transport is to measure the bulge; melting of the plume head often breaks through to erupt volcanically on the surface.

There are some forty known plumes causing hot spots on Earth. Small plumes are responsible for the Galápagos Islands in the Pacific and for Iceland on the mid-Atlantic ridge. One of the best-known larger ones has thrown up the Hawaiian Islands, a chain of giant volcanoes; the plume that made them has bowed up the floor of the Pacific in a dome a kilometer high. The Hawaiian Island chain curves westward; many of the "islands" do not rise above the surface, remaining as seamounts beneath the waves. Together, they reflect the somewhat curving movement of the Pacific plate, which rotates slightly over the plume, which remains fixed. The length of the Hawaiian chain times the rate of plate movement gives the age of the hot spot—approximately 120 million years; the plume that made Hawaii departed the core-mantle boundary in the early Cretaceous period. If all the islands and seamounts were piled one on top of the other, like so many Ossas on so many Pelions, the resulting mound would approximate in size Olympus Mons on Mars, the largest volcanic edifice in the solar system. The reason that Olympus Mons, which presumably was thrown up over a Martian hot spot, grew so big, many planetary geologists believe, is that Mars has no plate tectonics, and therefore the same part of the lithosphere always remained over the hot spot. Ossas on Pelions beyond count. The comparative planetologists on the Magellan team liked to point out that while there is nothing on Venus as large as Olympus Mons, neither are there any Hawaiian chains.

P LUMES may have to do with the way Venus loses its internal heat. All planetary bodies have heat to get rid of, even the icy moons of the outer planets. The heat is either left over from the process of accretion from smaller bodies from which the planets and moons were assembled in the first hundred million years or so of the solar system's life, or is still being generated by radioactive elements deep in the core, the mantle, or the crust.

There are three basic ways a planetary body gives up its heat: by conduction, which simply means the heat moving outward through the planet until it radiates into space, and which is the way most small bodies eliminate heat; by volcanism; or by crustal spreading and plate tectonics. The earth uses all three methods, though crustal spreading and plate tectonics are by far the most important, accounting for sixty-two percent of our heat loss. Plate tectonics cools the earth in two ways: the hot mantle material welling up at the rift cools as it spreads sideways laterally, like hot soup in a saucer. After subduction the sinking plates of cold lithosphere cool the mantle like lumps of ice dropped in hot soup. The earth is the only planet that is known to employ these methods of cooling.

Before Magellan arrived at Venus, there was a considerable dispute among the scientists regarding which of these methods, or which combination of them, Venus used to eliminate its heat. All three methods had their partisans, though the two main contenders were plate tectonics and conduction. Volcanism was dropping by the wayside largely because of Shoemaker's figure for the average age of the surface, 400 million years. If volcanism played a major role in heat loss, comparable, say, to the role of plate tectonics on Earth, the surface of Venus would have to produce a layer of lava perhaps up to fifty kilometers deep every hundred million years—and yet there was all that 400-million-year-old surface, with some of it as much as a billion years old. On Venus, despite all the signs of volcanic activity over the ages, the yearly average amount of lava added to its surface appeared to be only two cubic kilometers a year, or a tenth the amount for the earth. If these estimates were correct, volcanism could account for only one or two percent of Venus's heat loss.

Plate tectonics (as distinct from spreading, Head's main interest) was not really in the running, either. One major strike against it was the lack of water on Venus, which made a mushy, lubricating asthenosphere unlikely. Ironically, McKenzie was the most outspoken against the process on Venus. To Head's suggestion to me earlier that a waterless Venus might provide a glimpse of what our ocean floors would look like if the oceans were drained, McKenzie gave short shrift. "If you pulled the plug on our oceans, you'd see things, like linear features, running across the ocean floor—the transform faults or the boundaries of plates," he told me later. "You just don't see these things on Venus—I don't see them on Pioneer–Venus or Arecibo imagery, I don't see them on Venera—and so far I haven't seen them on Magellan imagery. I don't see the Freyja escarpment as a trench or possible subduction zone, and I don't see Aphrodite as a mid-ocean ridge. [Head had suggested both possibilities.] All these ideas about plate tectonics which I and my colleagues thought of for Earth are being transported by others to Venus—but I don't see any evidence for them on Venus at all! Maybe I'm getting old. The impression I have is not the similarity of Venus and Earth, but the differences. Of course, it's much easier to say that Venus is like Earth. But it doesn't look much like Earth to me. That's disheartening—it means we all have a lot more work to do!"

McKenzie argued that even if by some mischance there was plate tectonics on Venus, it couldn't be much of a source of heat loss, again because of Shoemaker's 400-million-year average age for the surface. "At that rate, you are losing hardly any heat from the interior by plate formation—assuming there is plate formation at all," he said, leaving the impression that plate tectonics that sluggish, about a seventh the rate on Earth, would hardly be worthy of the name.

If only by the process of elimination, the scientists were increasingly receptive to the idea that the chief method by which Venus lost heat and through which its surface was shaped was conduction—the simple transfer of heat outward through a fixed material. A more active form of heat transport, as Phillips had indicated with his analogy to a pot of water on a stove, is convection—the plumes, although most of the heat from plumes escapes through the lithosphere by conduction. With respect to heat loss, plumes had one drawback: on Earth, they are responsible for only six percent of the loss. However, they could play a larger role on Venus, where the other forms of heat loss are less efficient.

The manner in which a body loses heat largely determines how its surface looks. Phillips told me that he and others thought plumes were better than spreading or plate tectonics as an explanation of many of the features on Venus—in particular, the circular ones; they even see circles where others had not seen them before. They see western Aphrodite, the main body of the scorpion, as being composed of four giant circular regions—Ovda and Thetis Regio and Artemis, a huge circular corona to the south of Thetis, and also an unnamed circular region west of Ovda forming the scorpion's pincers. Phillips thought that the different stages in the life cycle of plumes explained some of the differences he saw in the circular features. He pointed at Beta Regio, well to the west of the pincers, which looked to him like the product of a fairly young plume—as the top of the plume flattens against the bottom of the lithosphere, a dome forms, and lava pours out of cracks in its top and sides. This was true of part of Beta. Then, as the plume gets older and stronger, the dome lifts up higher as the crust is thickened by massive intrusion of magma generated at the top of the plume, and it may try to collapse to the side. In some cases, the dome might begin to split, like an overcooked frankfurter, with one or more troughs or chasms, called graben, running down the middle. In some of the larger troughs, they are rift valleys whose sides have moved a few kilometers, or even tens of kilometers, apart, as a result of the lateral motion of the top of the plume spreading mushroom-like underneath the lithosphere. A rift valley runs through another part of Beta. Graben or rift valleys run through much of the equatorial highlands, too, such as Ovda Regio and Thetis Regio, in the upper body of the Aphrodite scorpion. Then, when the plume gets old and dies down, the dome collapses and the crust on top cracks and crumples. Possibly, Phillips believes, this is the origin of much of the tessera, which Head had suggested might be the result of compression due to spreading. When the plume dies away altogether, the dome subsides farther, leaving a vanishing hint of circularity, like the smile of the Cheshire Cat—the open pincers of the scorpion, for example, or certain arc-like chasms.

Although there are profound differences between the plume theory and the theory of plate tectonics, they are by no means mutually exclusive, as can be seen on Earth, where the two co-exist—the Hawaiian Island chain is proof. Head clearly recognized the likelihood of plumes on Venus and indeed made use of them to explain the mountainous Aphrodite and Beta regions, as Phillips did. Phillips recognized that there was rifting and horizontal movement of a few kilometers to a few tens of kilometers, though he did not attribute it to spreading, as happens on Earth at the mid-Atlantic ridge. Spreading implies horizontal motion of thousands of kilometers; rifting need not. (Despite the fact that a rift runs down the center of the mid-Atlantic ridge, spreading is what goes on there; the central rift is secondary to large-scale spreading. Confusingly, the Atlantic Ocean may have started out as a rift which then spread, and spread, and spread.) Where Head sees mid-Atlantic ridges and spreading on a large scale, Phillips sees rifting and splits in hot dogs—splits which in some cases had pulled apart a little and were filled with fresh lava. Indeed, the idea of spreading was principally what separated Head from Phillips, and what made Head's theory testable—the quality that Phillips respected. The test would come at Aphrodite.

I WENT out to Pasadena two months later, in the middle of November. In the interval, the spacecraft had suffered no more LOSes, though the alert signals caused by the stuck bits and the faulty addresses in memory B, which had startled the engineers immediately after the rocket had been separated following the VOI burn, had occurred twice more, briefly in October and more seriously in early November, when there was a series of fifty-seven alerts. Memory B had been off line ever since the August episode (with the exception of the times it had inadvertently been brought on line during the LOSes), and consequently these alerts had no ill effects. Had it been on line, the alerts very likely would have precipitated RPEs and heartbeat failures. No one yet knew what caused the stuck bit and the wrong addresses, but there were a number of theories, most of which centered on the separation of the rocket motor. The theory that Spear and a number of other engineers favored was that the electrical capacitor for firing the explosive bolts had shorted out to the chassis of the spacecraft, and this in turn had damaged a silicon chip—which presumably had already been weakened, possibly by solar radiation during the trip from Earth. Many of the flight engineers hoped the chip would heal itself, which frequently happens. But they did not know when, or even whether, this would happen; or indeed if the theory was right in the first place.

With spaceships, as with other kinds of ships, there are always small problems to occupy mariners. One day, the solar panels began oscillating during mapping. The engineers were baffled until they realized that for a time the attitude of the spacecraft was such that the panels were edge-on to the sun, and the sensors meant to orient the panels broadside to the sun were having a sort of nervous collapse. The problem cleared up a few orbits later. On another day, an orbit's data were lost when the spacecraft computer erroneously accepted a bad star-cal update. Problems happened on the ground as well as in space. On a third day, three orbits' data were lost when a pair of redundant fuel pumps for a pair of redundant generators failed in Madrid. The generators ran out of gas. The flight controllers thought the spacecraft had taken a siesta. On a fourth day, all the pictures processed at the radar-imaging laboratory came out blank; it turned out the problem had to do with what computer technicians call slipped bits, bits that had been misplaced in the lab. When they were restored, the pictures were fine—though the time it took to solve the problem, several weeks, put the image processing behind schedule.

The most serious problem, though, and the only one that hadn't corrected itself or been corrected, was one that developed in space gradually, over several days. One of the two redundant tape recorders aboard the spacecraft for storing the mapping data prior to transmitting it to the ground was apparently losing two of its four channels. For the present, the spacecraft was being programmed to avoid those two channels, while the contractor that made them, Odetics, Inc., in Anaheim, studied the problem; but there was some fear that the degradation might spread to the other channels.

The engineers had by no means forgotten the two LOSes, and at Denver Rick Kasuda, the leader of the attitude-control software team, and Bob Reilly, a software engineer under Kasuda, had been going over the delayed recorded data from the memories to find out what had caused them. Reilly had written some of the programs in the spacecraft computers several years earlier, before he had switched to another project; he was reassigned to Magellan after the second LOS, to help with the troubleshooting. He is a short, clean-cut man in his late twenties who, when I met him much later, was wearing gray pants and a white shirt—the uniform of computer engineers. Kasuda, who is about the same age, is taller and darker than Reilly, but like him talks with a brisk and rapid-fire logic.

"With the data from the memories, we started at the end of the problem and tried to work our way back—what we call backchaining," Reilly told me later. "The trouble with backchaining is that there are a great many pathways, and you might not go down the right one. You're always on the lookout for something that might make a hypothesis. You work by trial and error. But the evidence gets scarcer as you go back. You know the effect—but the evidence for the cause gets fainter and fainter." Reilly and Kasuda were given until the end of October to work on the problem. The end of October came and went, without an answer. Reilly and Kasuda went on to other tasks.

The scientists were full of sympathy and admiration for the engineers and the Herculean efforts they had made to keep the spacecraft going, and they went to some lengths to make them feel appreciated. Regularly, one or another of the scientists briefed the engineers about their discoveries on Venus. When Peter Ford, a tall, lanky Englishman with graying hair in a constant landslide down his forehead, who is a radar physicist at M.I.T. and the chief scientist for the altimeter, was briefing the scientists on the latest refinements of the data concerning the heights of the Danu and Vesta escarpment, he began: "First, on behalf of all the science team, I want to say thanks for all the sleep you've lost, keeping the spacecraft going. We really appreciate it." Indeed, the scientists were forever inviting the engineers into their area and plying them with the latest pictures—clearly, with the ulterior motive of reinforcing their enthusiasm for keeping the spacecraft going.

ON Monday, November 12, at eight in the morning, I arrived at the science room for a meeting of the science team. There weren't many scientists around yet. The walls were covered with F-BIDRs, and the overflow—rolled into tight little scrolls—was stored in rows of tiny cubbyholes similar to those in a Chinese scriptorium. At the moment, the flow of F-BIDRs had ceased, allowing the scientists to catch up on what had been accumulated—since the end of October, the spacecraft and the planet were behind the sun in relation to Earth, a situation called superior conjunction, which had started the last week in October and was about to end; new F-BIDRs should start coming down before the end of the week.

Jim Head was looking at a pile of large photographs, each a couple of feet square, which were mosaics of selected parts of the surface. The technical name for them was full-resolution mosaic image data record, but they were called F-MIDRs to rhyme with F-BIDRs, the noodles from which they were derived. They hadn't existed at the time of my last visit, because there hadn't been enough orbits—one F-MIDR could cover an area including parts of up to two hundred orbits. If you laid the highest-resolution F-MIDRs of the entire planet side by side, I was told later, they would make a rectangle 350 meters long and more than a kilometer wide—yet even at that scale you would be unable to see a single pixel or data bit ("pixel" is an abbreviation of "picture element"). "So, if you are wondering why the volume of data is so great and the data processing is so complicated—that's why," my informant told me. Eventually, F-MIDRs covering the entire planet would be put on over a hundred compact disks for computers; they would cost the scientists a dollar each, or about one 500-millionth of the cost of the mission they were the major product of.

I asked Head to give me his impression of the mosaics and other material that had arrived since my previous visit. "We've been getting great stuff!" Head said. "We always knew what hundred-meter resolution meant, but we had no idea how continually exciting this stuff would be. At first, it strikes you like a painting—it's beautiful. Then the delicacy of the geological structure just grabs you. And then, as you look at it, you are struck by the complexity of all the features you see. In the mountains, you can see all the hills and valleys and other structuring. Sometimes you focus on the faults. Sometimes you see so many details that you have to step back to see the picture at lower resolution, so you don't miss the forest for the trees. But the detail is what strikes you—you can actually see the talus moving downslope, or signs of the dust being moved around by the wind. There's not as much dust as on the moon or Mars—the very dense atmosphere heats up and burns the smaller asteroids, so you don't get the fine dust of those places. On Earth—or the moon or Mars, for that matter—you can wander all day without finding an outcrop of bedrock, because it's all covered up. But on Venus, where there is no water and so much less dust, all the bedrock is exposed. The entire planet is like a great big road cut. So the exciting thing is that what we're looking at is total outcrop, total exposure of everything that's happened to it. On Earth, as soon as a mountain or a canyon is formed, you get erosion, and soon you lose the evidence of what happened in its first formation. So the really exciting thing on Venus is that we have laid out for us the complete sequence of formation of, say, mountain ranges. In each, there is a complete record, and the challenge is to sort out what is a very complex record. But it's all there. So at a glance you can see the compression faults which threw the mountains up, the increased faulting as they were thrust higher, the extension faults which later pulled them down, and the troughs and graben that were part of their collapse. One exciting thing is that the topography often is so steep that it forces you to think that many of the mountains are forming today—otherwise, they would long since have subsided and collapsed. So something is pushing the mountains up."

I asked Head if he was still holding the door open to spreading and even to plate tectonics as a possible force for mountain building. "It's too early to tell," he said. "We haven't even gotten to Aphrodite yet."

At last Saunders called the meeting to order, and the scientists took their seats around the horseshoe of tables. At the front of the room, propped up on an easel, was a picture of western Ishtar, and in the distance Danu Montes, crowned with white as though snow-capped, rising one and a half kilometers above Lakshmi Planum to the right and four and a half kilometers over the lowlands to the left. The picture, in hot Venus orange, was in perspective, as if the viewer were standing on the ground, and its affect was three-dimensional. Unfortunately, the view was broken in three places by undulating black ribbons —the gaps left when Madrid ran out of gas. Computer programmers at J.P.L.'s image-processing laboratory, in collaboration with geologists with the U.S.G.S. at Flagstaff, had been able to combine Magellan's altimeter data with the SAR data to create the effect of being on the surface. To give a sense of the relief of the terrain, the geologists had exaggerated heights by a factor of 22.5; the effect was dramatic. Geologists sometimes distort the relief of images in order to bring out things that would not otherwise be noticeable. Looking at Danu rising spectacularly on the horizon in the picture was a little like looking at a distant mountain through a pair of binoculars—a bump on the horizon becomes a jagged peak. However, the vertical exaggeration did rather more than bring the mountain closer; it turned gradual slopes into precipices. And because of the heat and weakness of the rocks, most mountains on Venus have the profile of gentle hills. Its mountains look like molehills, even if the moles are titanic ones. Even though their peaks rival the Himalayas, they don't tower majestically; they rise, like a sort of giant groundswell. However, the pictures gave the public the impression that Venus was more scenic than it really is.

Since my last visit, the scientists had quite clearly moved from the stage where they were pointing at features and arguing over what they were to a new stage where they had enough information so that they could begin to make some generalizations about what they were seeing. The first speaker was Peter Ford, who reported on the latest altimetry data. A curious phenomenon, Ford pointed out, was that the upper parts of Danu and of several other high mountains, including two recently imaged volcanoes to the west and north of the scorpion's pincers, Sif and Gula Mons, 2.8 kilometers and three kilometers high, became more reflective on radar at about the 2.5-kilometer altitude, so that they appeared snow-capped. Sif had a white top; on Gula, two-tenths of a kilometer higher than Sif, the whiteness reverted to the orange that the rest of the landscape had been colored. On Maxwell, the Everest of Venus, for some reason the whiteness did not start until the 4.7-kilometer level. The phenomenon had first been noticed by Pettengill and Ford in the Pioneer–Venus imagery more than a decade earlier, but they were seeing it in the Magellan imagery with greater precision and in more places. Ford thought the white stuff, as the reflective material was called since the scientists did not know what it really was, had to do with a chemical change in an unidentified material that was sensitive to the temperatures and pressures of that particular altitude. The best anyone could come up with, he said, was an iron oxide in the rocks that under those conditions turned into a more radar-reflective magnetite.

Ford was followed by Jerry Schaber, who had been counting craters and now reported that on the fifteen percent of the planet mapped before superior conjunction, he had found 119 craters up to a hundred kilometers in size—from which he predicted that between 700 and 800 craters would be found on the planet as a whole. (As it turned out, his estimate was only a little low.) He was surprised at how young most of them looked, he told me later—in most cases, even the ones that didn't have the fresh, reflective ejecta blankets, like Danilova, Aglaonice, and Saskia, still looked sharp and pristine. Unlike the moon or Mars, Venus has no craters that were degraded or subdued, disappearing into the ground like ghosts. In part, Schaber felt, this was because of the thick atmosphere, which protected the surface of Venus from micrometeorites responsible for much of the degradation of craters on the moon; moreover (as we have seen), there was relatively little dust to be blown into craters, further subduing their contours —at least, none of the craters seemed covered by dust. In addition, of course, the Venusian craters *were* young, as craters go—on the average, 400 million years old, if Shoemaker was right. On the moon, where craters can be billions of years old, the ones that are a few hundred million years old seem born yesterday. The relatively young age of the Venusian craters was a puzzle. "If the rest of the planet looks like the fifteen percent we now see, we have a problem," Saunders told me later. "If it looks like this, we will have to revise some of our thinking."

The meeting broke up early. A little later, when I talked to Schaber in his cubicle at the rear of the back room, he looked peeved and exhausted. For one thing, Pettengill (a physicist) had asked Phillips (a geophysicist), not Schaber (a geologist), to write the section on impact craters (normally, the preserve of geologists) for an official report to NASA (known euphemistically as the forty-five-day report, even though sixty days had passed since mapping started and the report wouldn't be completed for several months). That the geophysicist head of the interior-processes working group (Phillips) was to write the section on cratering in the forty-five-day report, instead of the head of the impact-processes working group (Schaber), was something no self-respecting geologist could accept lying down. I found out later that the geophysicists had convinced Pettengill that impacts, given the tremendous amounts of energy released and the effects of the atmosphere on them, were dynamic events and hence the preserve of physics. However, I also got the impression from some of the geologists that they felt the geophysicists had a sense of intellectual superiority to the geologists and therefore of entitlement—the way, perhaps, European colonists felt when they found bothersome aborigines occupying a particularly alluring bit of terrain. And their solution to the problem, these geologists implied, was the same.

As if all this wasn't enough, Schaber was also upset because, the night before, he had found several more F-BIDRs he hadn't known about, rolled up in the tiny pigeonholes, and had been up much of the night counting the craters on them. He wasn't through yet, though already it was clear he would have to revise his estimates upward. Unlike many of the other team members, Schaber had no graduate students to do the drudge work, for the astrogeology branch of the U.S.G.S. was not a teaching institution. Conse-

quently—unless he could co-opt someone else's graduate students or some friends—he had to find and
count every single crater on Venus with his own
fingers. (He admitted to being envious of some of his
colleagues, like Jim Head, who had a dozen students
helping him.) Counting the craters was particularly
important now, Schaber felt, because even before
Saunders he had sensed that the high proportion of
relatively young craters would require a considerable
revision in the thinking about Venus—a revision of
certain theories of the geophysicists about the interior
processes of a planet. Accurate crater counts were
part of the required evidence.

For the time being, he had put the crater counts
aside. He pulled out a big mosaicked picture and
showed me a huge crater filled with black lava,
Cleopatra, tipped at a crazy angle on the side of
Maxwell. Lava-like flows poured down the crater's
eastern flank, inundating the valleys and plains far
below—an observation first reported by Schaber and
two colleagues in 1987 after photogeological analysis
of the Venera 15 and 16 images. In the imagery, the
flows looked black—which in radar, of course, did
not mean they were black, but smooth. As impact
craters get larger in diameter, their profiles get flatter,
and the largest ones can be surrounded by one or
more rings of steep mountains thrown up by the
impact. Cleopatra is about a hundred kilometers
across and two and a half kilometers deep. It has two
concentric rings of mountains, the inner one (which
contains the blackest—or, rather, the smoothest—
material) perhaps sixty kilometers across. Though it
looked like an impact crater, there had been some

doubt whether it might not in fact be a volcanic crater—the same sort of argument I had heard in the first few days of mapping. Venera data had not revealed a hummocky rim such as impact craters normally have, and it had indicated that the crater bottom of the inner ring was perhaps a kilometer deeper than the outer ring—and this would only be possible in the case of a volcanic caldera, which slumps in the center. "In 1987, I wrote a paper saying it may be volcanic," Schaber told me. "However, the Soviet scientists—in particular, Basilevsky and Ivanov of the Venera 15 and 16 team—thought it was impact. In their Venera imagery, it seemed to have a bright aureole around it, suggesting an ejecta blanket, and that's what persuaded them. I wasn't sure. Now, just today, the Magellan data have come in, and the hummocky rim and the ejecta blanket are very clear. What's more, the Magellan altimeter indicates that the floor of the inner ring and the floor of the outer ring are at the same level—in other words, the entire crater floor is flat, the way you would expect with a large impact crater. So I have no trouble now with the idea that Cleopatra was made by an impact.

"Another question is, if all this black lava was melted by impact, how did it get through both the inner rim and the outer rim of the crater to go downhill? You can see that the flow of lava starts on the inner floor, crosses the outer floor, and cuts a canyon through the entire thickness of the eastern rim wall before plunging into the valley below. I talked to Shoemaker when I was in Flagstaff last week, and he suggested that if Cleopatra was an impact crater it would have had a central peak which is now gone—the peak probably slumped back—and if it did that while the lava was still liquid, it would have forced the lava to break through the mountain barriers and run on downhill. But that doesn't account for why there is so much of it.

"Some of us feel that there is too much lava in the valley below to be explained by the heat of the impact melting the rocks. Randy Kirk, a colleague of mine at Flagstaff, figured that if you scooped it all up and poured it back in, it would fill the crater two-thirds full. So we feel that the impact may also have induced real volcanism. Basileveky and Ivanov and Kirk and I are collaborating on a paper that combines all these ideas about Cleopatra." Increasingly, I was finding on Venus that more than one hypothesis can be correct.

Several months later, it turned out that there had been an error in the processing of Magellan's altimeter data, which had to be reprocessed; and when Kirk at Flagstaff analyzed the new data in his computer, he found that the central part of the crazily tipped crater was indeed deeper than the outer part. This pushed him and Schaber back to the original Venera estimate that the center was as much as a kilometer deeper. Because of the ejecta blanket, though, Schaber held firm to the idea that Cleopatra was basically an impact crater. He and his collaborators were then faced with reconciling their combined hypothesis with the revised altimeter data. "If the inner floor is a kilometer lower than the outer floor, then it must have subsided sometime after the flows poured out," he said. "So there is still a mystery to be solved about Cleopatra, and our paper will describe the options."

I gave him an option for getting the lava out of the now-deeper crater and into the valley below: Perhaps Maxwell's flank had been steeper when Cleopatra was formed, so that the crater then was even tippier than it is today and the melt poured out easily? And then Maxwell's flank had become shallower under the influence of crustal flow, so that Cleopatra became less tippy, until it reached its present angle —thus giving the appearance of a problem where there in fact was none? (It is impossible to be around planetary scientists for very long without wanting to participate in their games.) Schaber dismissed this option as unworthy of even the rankest geological patzer. If the mountainside had flowed, he said, the crater would have, too, so that it would be distorted, as if it were a crater in a painting by Dali—but there it was, fresh and round, an eyeball peering enigmatically from the side of the mountain, sort of like the one on the pyramid on the dollar bill. Clearly, Cleopatra was comparatively young, even in relation to the relatively youthful mountain.

THE scientists were beginning to find some strange features on Venus—in particular, ones they called splotches, horseshoes, and pancakes. All planets and moons visited by spacecraft have their own curiosities, Earth most of all. A Venusian or a Martian observing our planet would be struck first by our biosphere (an outgrowth of the geology of our planet) and then by other geological oddities such as the granitic continents, the oceans, lakes, rivers and waterfalls, and— most wonderful to an extraterrestrial—the only terrain in the solar system (the quirky, sulphurous Io aside) mostly devoid of impact craters. Compared to these, the Venusian curiosities are quite modest:

THE SPLOTCHES. Out in the main room, Laurence A. Soderblom, a geologist with the U.S.G.S. at Flagstaff, was studying an enlargement of several sections of mosaic spread on one of the big gray tables. Soderblom, the leader of the satellite group for the science team of the Voyager mission, was something of a specialist on the strange and bizarre. He had, for example, developed a convincing theory to explain the geysers on Neptune's icy moon, Triton, whose surface temperature, thirty-eight degrees above absolute zero, at the opposite extreme from Venus's, had not led the scientists to expect a phenomenon normally associated with volcanism. On Venus, he pointed out several small craters under twelve or fifteen kilometers in diameter, each of which appeared in the middle of an irregular but vaguely circular black spot between fifteen and twenty kilometers in diameter. Elsewhere, he indicated several more splotches of the same size minus the crater. He and others had puzzled over these spots for some time, and now he said he thought he knew what they were.

"The thickest part of Venus's atmosphere can be thought of as a layer twenty kilometers thick," Soderblom said in a voice that conveys information with the force and rapidity of a geyser—and with something of its enthusiastic impulsiveness. "If you collapse it fortyfold, you wind up with a half-kilometer layer that has the density of rock. At Venus, you can think of a comet or asteroid as impacting in the atmosphere—and some make it to the surface and some don't. In other words, the surface is armored against any asteroid or comet that is below a certain size, which is easy to compute. A stony asteroid in Venus's atmosphere will slow down roughly by a factor of two every time it travels fifty times its own length. So what is the smallest one that will make it to the ground? An object a few hundred meters in diameter will just barely make it to the surface—that's why the smallest craters, we are now finding, are a little under two kilometers, the size that such an object would make." The best-known crater on Earth, Meteor Crater in Arizona, 1.2 kilometers in diameter, was made by a body approximately fifty meters across; it would not have made it to the ground on Venus—though, according to Shoemaker's flux, there would be 88,000 craters that size on the planet if it lacked an atmosphere.

I said I was not clear how all this related to the splotches. "When an asteroid or comet hits the atmosphere, it sets up shock waves which precede it to the ground," he said. "Regardless of whether the body burns up or not, the pressure will hit the ground—what we call a ground slap. For a hundred-meter asteroid, just large enough to make it to the surface, the slap hits the ground with a force of ten million joules per meter squared—if someone really swings a sledgehammer, the head will hit the ground with only ten thousand joules. And this force is enough to pulverize the rock in the upper couple of meters of the surface; it turns the rock into something the consistency of flour, and this shows up dark on the radar—hence, the splotches. For impacts this size and larger, the splotches are all about fifty kilometers in diameter because that's a few times the rough height of the atmosphere, and they can spread out only a few times that height. Much larger impacts often have no splotch surrounding their craters—the ejecta from the impact probably obliterated it. Smaller bodies that impact have these black halos, down to fifteen kilometers in diameter. Asteroids that don't make it to the ground at all just leave the splotches—sort of ghosts. And asteroids smaller than perhaps fifty kilometers don't leave any trace at all."

Soderblom is the kind of fellow who, when he gets an idea, extracts from it every bit of juice that is there—an unstoppable flood of ramifications and new wrinkles. "If the theory is correct, you ought to be able to make predictions from it," he said. "For example, there should be fewer ground slaps on higher terrain because the atmosphere is thinner—and, indeed, there are no splotches around the craters on the Lakshmi plateau or Maxwell Mons." Furthermore, if the theory were correct, it could be used to determine other matters—for example, if the splotches' diameters were a function of the depth of the atmosphere, scientists should see if older splotches had the same width in relation to their craters as more recent ones, to see if the atmosphere had the same thickness through time. Or it might be possible statistically to determine the percentage of impacting bodies that were asteroidal as opposed to those that were cometary—comets being the only other bodies responsible for impacts. Comets, which are snowballs of ice and carbonaceous rock, impact in the atmosphere with enormous energy but disintegrate more rapidly than asteroids, which are solid bodies of iron or stone; and consequently the larger splotches with the smaller craters might be cometary.

I asked Soderblom if ground slaps had been detected on other planets with appreciable atmospheres, such as Earth. He said one example was the Tunguska event in northern Siberia in 1906, when a comet is thought to have detonated in Earth's atmosphere; trees were knocked down out to a radius of fifty miles. The detonation might have been the ground slap, and the ring of fallen trees a woodsy version of a splotch. He thought splotches should be looked for on Mars as well as Earth. A few months later, Peter Schultz, the expert on impacts from Brown, told me that he thought he had found evidence for ground slaps on Mars, and also on a recently discovered, fairly fresh, impact crater on Earth. And Shoemaker, who had been the first to prove that Meteor Crater in Arizona was of impact origin, told me that several decades ago, when he had been excavating underneath the crater's ejecta blanket to find plant material (he needed it for carbon dating, to find the date of the impact), he had found none whatsoever. "Now I suspect all the plant life was crushed and pulverized by the shock wave," he said.

To Shoemaker and many others, the halos on Venus—and the idea of the ground slap—were a complete surprise; they were, moreover, one of the first examples of a discovery on Venus that could be used to make discoveries on other planets. The reason they were found on Venus first was not only that the density of the atmosphere resulted in a far more pronounced splotch but also that on Mars and Earth the far fainter splotches were quickly covered over by windblown deposits or (in the case of Earth) washed away or covered over by water. On Venus, the splotches last longer because there is no water and only minimal surface winds.

THE HORSESHOES. I found Arvidson and Wood in the first cubicle, poring over a picture of a large crater, Aurelia, that had a faint black splotch partly surrounding it; only, instead of being a circular splotch such as the ones Soderblom had shown me, it was a giant parabola that opened out in two arms, extending about a thousand kilometers to the west— it looked like a horseshoe hanging on a nail, the nail being the crater. "Larry's pressure-wave theory explains his splotches, but it doesn't explain these," Arvidson said. "We've only found two or three of them. They all open to the west. Is it an impact phenomenon or is it a wind pattern?"

"It could be both," Wood told me. "An impact throws stuff up in the air, and the wind blows it."

Arvidson, who agreed with this proposition, said that one difficulty was that the winds near the ground were known to be slow. "One model is that if you throw the stuff up high enough, it will get caught in the fast prevailing westerlies, which were discovered by a number of Soviet and American probes, in particular Pioneer 13, which in 1978 dropped four probes in the atmosphere," he said. In June 1985, the Soviet Venera spacecraft deployed two balloons in the atmosphere at an altitude of about fifty-four kilometers. The winds blew them westward at about 250 kilometers per hour, each traveling more than 11,000 kilometers during the forty-six hours they were monitored. All this, he thought, fitted in with the fact that the horseshoes found so far all surrounded bigger craters—the impact had to be sufficiently large to get the dust high enough to reach the westerlies. Furthermore, they appeared to be connected only with the youngest large craters, such as Aurelia—the debris scattered a thousand kilometers across the surface was too thin to survive for long, even in Venus's quiet lower atmosphere. (Aurelia was the name of Julius Caesar's mother—and also, a well-informed geologist told me later, of Arnold Schwarzenegger's.) In the next few months, Arvidson found a few more horseshoes—one, he told me when I saw him again in February, was draped over the heights of Gula Mons. That particular horseshoe went over some flows surrounding Gula—the echoes from the radar were less bright because the dust made a blanket smoother than the flow. But the echoes from the underlying flows could be detected, too, showing that the horseshoe material was too thin to bury it completely—at the most, the deposit was a meter thick. Later he found that thirty-five percent of the plains gave off the radar signature connected with the horseshoe debris. This, of course, meant that the deposit had to be far thinner than that—just a thin coating. Because of their association with the youngest craters and because of their farflung nature, Arvidson believed the horseshoes would be useful for dating much of Venus's surface.

The sluggish winds near the surface had left other much less obvious patterns called wind streaks which Arvidson had also been studying. To him, they are like arrows on a weather map, indicating wind direction and also speed. Arvidson told me that a member of his group, Ronald Greeley of Arizona State University, had measured the wind streaks on Venus and had found that there is a global pattern, which would not be the case if volcanism or impacts alone were responsible for them. The streaks south of the equator point north, and those north of the equator point south—the Venusian version, Greeley had said, of what on Earth is called a Hadley cell, which consists of easterly and westerly bands of winds, driven horizontally to the equator by the Coriolis force of the earth's rotation. Venus rotates so slowly that its Coriolis force is negligible; its north-south winds are what our winds would be like if Earth stopped rotating. Because they are powered by the sun's heat (only twenty percent of which reaches the surface), the Venusian winds reach their peak at noon of the eight-month day; judging by the streaks, they are very slow, no more than two or three miles an hour—not enough to relieve the torpid, sultry Venusian doldrums. "The lower atmosphere is so massive that it travels very slowly under normal conditions," Arvidson told me later. "You would never get gale-force winds on the surface of Venus, other than as a result of an eruption or an impact."

....

THE PANCAKES. In the main room, Saunders, John Wood, and Alexander Basilevsky were studying an F-MIDR mosaic that had just come from the processing lab. Basilevsky, the Venera 15 and 16 scientist who had worked on crater counts and Cleopatra among other things, was one of three Soviet members of the Magellan team. He is a geologist at the V. I. Vernadsky Institute of Geochemistry and Analytical Chemistry in Moscow, which had been in charge of the scientific investigation for most of the Venera missions.

The area the three scientists were looking at, just east of Alpha Regio and about thirty degrees below the equator, had been mapped only a week before, but the lab had rushed the special product through, at the urgent request of the scientists. At first I thought I was looking down at seven rocket bursts that had been set off in a long row, with some of them impinging on each other; or perhaps it was seven parachutists that had jumped seconds apart, their chutes mushrooming simultaneously. The image preferred by the science team was pancakes—"Not pancakes, but blini," Wood told Basilevsky, who happily appropriated the term—and indeed they looked exactly like seven overlapping dollops of batter spooned in a row on a hot griddle, turning solid before your eyes. Their tops looked blackened and cracked, the way pancakes do when they are first turned over, and the sides looked white, as if the new batter freshly turned was oozing outward. Some of them had little craters or pits in them—like bubbles that had burst in the batter. (Batter poured on the

....

surface of Venus, as hot as a griddle, would do all these things.)

The pancakes were twenty-five to thirty kilometers across and two hundred or three hundred meters high at the edges. Each one had what looked like a vent almost precisely in the center—suggesting that they were some sort of volcanic feature. Wood wondered if they might not be pyroclastic material that had erupted from the vent on a windless day, the ash and clinkers falling evenly all around. Saunders said he didn't think so. Pyroclastics tended to make conical structures that rose at the same angle from the surrounding plain to the apex—like Etna or Vesuvius; the pancakes were clearly pancakes, with flattish tops and steep edges. Saunders said that the most likely cause of the pancakes was a thick lava extruded fairly quickly from the central vent onto a very flat surface, so that it spread evenly—just the way batter does on a grill. The succession of pancakes reminded him of the row of islands and seamounts that make up the Hawaiian Island chain, and Saunders wondered if the pancakes might also be caused by crust moving across a stationary hot spot; however, he and most scientists came to the conclusion that the vents were lined up along a crack in the surface.

Liquid lava spreading from a vent, indeed, is the way similar features are formed on Earth—there are little pancakes (mini-blini or crepes, the scientists couldn't decide which to call them) in Siberia and in the western United States, but they are small, no more than a kilometer in diameter, and highly eroded. On Venus, of course, the grill is a lot hotter, and the lava stays fluid a lot longer, and flows farther.

Brennan Klose, Wood's graduate student, had seized on the pancakes early and planned to make a study of them—possibly for his doctorate. "I wrote my bachelor's thesis at Harvard predicting that there would be pancakes on Venus—and boom! there they are!" he told me. Later I asked him what had made him predict in his senior year at Harvard that there might be pancakes on Venus, and learned that it had to do with the fact that the smaller, more eroded pancakes on Earth—the best examples in the United States are a batch near Taos, New Mexico—were of a highly evolved, granitic material called rhyolite. Lavas that form ordinary granite, the stuff of continents, cool slowly in huge underground reservoirs, resulting in relatively large crystals, but rhyolite results when the same lavas are erupted onto the surface, where they arrive in smaller quantities and cool more rapidly. Rhyolite and other granitic melts are normal products of volcanoes overlying subduction zones, such as Mount St. Helens, and they contain a lot of subducted water and other volatiles. "My argument was that the granitic lava is very viscous, but it also contains a lot of bubbles (the water and volatiles), which explode when it is vented rapidly onto the surface as rhyolite—which on Earth is often associated with pyroclastic eruptions," Klose said. "Look at eruptions like Mount St. Helens, which not only produce rhyolite but are very explosive, with ash strewn all around! However, I reasoned that if there were any rhyolite or other granitic melts erupting on Venus, where the surface pressure is so much greater, there was a good chance any bubbles would be held in—all the lava on Venus would be like champagne in a corked bottle which keeps it under pressure and bubbleless. Consequently, such eruptions on Venus would be much less explosive, and the lava's viscosity would be more apt to result in pancakes—like the ones on Earth, but larger. And I was right!"

Klose told me later that the perimeter of the pancake was probably determined by the hardening of the front of the circular flow, as it would have been the first part to cool; the remainder of the lava then may have filled up the interior of the circle—indeed, pancakes on a stove do something of the sort, when they spread as far as they are going to. Though the pancakes looked as if they were higher toward the center, where the vent was, Peter Ford later suggested that this might not be the case. When the altimeter—its ground tracks trailing a few days behind the SAR's because it was looking straight down whereas the SAR was looking to the side—finally passed over the pancakes, Ford announced that their centers appeared to be flush with the ground level of the surrounding plains. He speculated that when the central vent had stopped emitting lava, some of it had flowed back down whence it had come—leaving behind a honeycomb. On Earth, the central parts of the pancakes are quite porous. Either way, the altimeter would see right through to the ground. In Ford's opinion, the pancakes weren't pancakes at all, but English muffins, which (when cut in half and toasted) are also low and frothy in the middle. However, the name "pancakes" stuck—and indeed Saunders and some other geologists were skeptical of what they regarded as a radar physicist's foray onto their terrain.

Saunders told Pettengill, the principal investigator of the radar experiment, that he thought the data was in error.

The altimeter data was indeed in error. The following May, Ford discovered there had been a mathematical problem in processing it—the same problem that had thrown off the estimates of the depth of the center of Cleopatra. Each measurement the altimeter makes, which it does every four or five kilometers along its ground track, consists of information from several bursts of radar which are combined in the processing laboratory at J.P.L. For a particular algorithm (or series of formulas) in the software that does the job, physicists use a minus sign where engineers use a plus sign. Ford had made sure that the physicist's method was used throughout. However, late in the game a new, faster processor was acquired; no one noticed that its algorithms followed the engineers' method and that pluses and minuses were reversed. As a result, bursts of altitude data from different points got mixed up—and hence the data thought to be from the center of the pancake had in fact been taken from the floor of the plain around it. "I and a graduate student discovered the error at two o'clock one morning at the end of May," Ford told me later. "We were trying to refine the data for Dan McKenzie, who was getting interested in the pancakes and who was coming over from England the next day to look at the data, and we were pushing our processing higher and higher for him. We were working with a pancake that appeared perfectly round on the radar imagery, but the enhanced altimetry measurements turned out to be all scrappy. I said, 'Oh, shit!' which is what people say when they discover a major error at two in the morning." The error was present in all the altimetry measurements done since January, covering more than half the planet. "In a situation like this, the only thing you can do is come totally clean and notify all your colleagues immediately. You have to tell the people investigating pancakes and other features affected by the data, like craters or the chasms, to stop working. Then you do everything in your power to fix the problem," Ford said. As a result of the fixing, the pancakes reacquired their pancake status—and (as a result of Kirk's interpretation of Ford's new data) Cleopatra its central well.

Scientifically speaking, planetary missions such as Magellan are tumultuous affairs. They condense into a very few years an exponential increase in the information about a new planet equivalent to the increase in our understanding of Earth which has taken place over the last two thousand years. In that period, innumerable errors in our understanding of Earth—unavoidable and perhaps even necessary in the development of science—have cropped up, and some of the greatest names in science have been associated with correcting them: Copernicus, who replaced a geocentric solar system with a solarcentric one; Charles Lyell, who first indicated that Earth was not a few thousand years old but far older. On this scale, Ford's error with the altimeter data and the muffin-like pancakes was a minor blip—in terrestrial terms analogous, perhaps, to the misconception based on squashy data explorers picked up from the Eskimos that there was a navigable Northwest Passage in the ocean at the top of Canada, and which took about a century to correct. Ford and his associates took just two months to correct the altimeter data and issue new measurements, working pretty much around the clock. During a planetary mission, there is enormous pressure on scientists like Ford to get the basic data out because other scientists need it for their work, and there is just as great pressure on these other scientists to interpret it. The situation, especially in the early stages, is fraught with the potential for errors, and they happen. As with the altimeter data and the pancakes, they are corrected as fast as possible. Errors such as this are annoying to the scientists, but they are accepted as part of the scientific process—at least, as long as that process includes human beings. "Mea culpa," Ford told me contritely. His sin—if there was one—had long since been forgiven by his colleagues and the episode forgotten. Scientists are generous to their colleagues who admit errors quickly—and merciless to those who don't. They agreed with something else that Ford said: "What really matters in a planetary mission is to get it right." He made sure that he did.

BASILEVSKY had first been shown the mosaic of the pancakes by Klose, and the two of them had looked at it together. (At Basilevsky's invitation, Klose would spend the following academic year, 1991–92, at the Vernadsky, in a friendship nurtured by the pancakes.) The pancakes were a surprise to Basilevsky, as they had not been evident on the Venera imagery—though when he got back to Moscow and searched the Venera images in the light of his new knowledge, he found small circles that he had not paid attention to before at the sites where the pancakes were, and he was later able to inform his Magellan colleagues that they should expect to find over a hundred of them on Venus. Basilevsky is a wiry man with a bright face who had been on the science teams of the Soviet Lunakhods (the automated cars that drove around the moon), of six of the seven successful landers on Venus, of the two Venera radar mappers, and of the ill-fated Phobos mission. He is one of the earliest links between Soviet and American planetary scientists, an association that goes back to a friendship he struck up with Harold Masursky in 1972. In the last decade, he has formed a close association with Head. Such friendships have flourished in periods of détente, and they flourished even more in periods of increased strain between the two countries. Between 1982 and 1987, when the Soviet–American space treaty had lapsed, relations between the two space programs were largely kept alive by unofficial contacts between scientists. A major forum for these contacts was a series of biannual symposiums sponsored jointly by the Vernadsky Institute and Brown University, which Basilevsky and Head have largely run. I was present at one in Providence in March 1989, when Basilevsky presented Head with the first complete set of maps of Venus based on the Venera 15 and 16 imagery, which he told Head he hoped would be useful for Magellan. (The Venera imagery was part of a trade in which the Russians received some of our Viking imagery, useful to them for their missions to Mars in 1994 and 1996.)

One of the first things Basilevsky did after he saw the mosaic was to look at the Magellan imagery of the area where the first Venera lander, Venera 8, had touched down, an ellipse of several thousand square kilometers. He discovered a pancake in the ellipse. Although it was by no means as perfect as the ones in the mosaic, it was a pancake nonetheless— one that had perhaps been nibbled around the edges. The discovery was not purely serendipitous. All the landers had been able to do chemical analysis of the soil, and all of them, except Venera 8, had indicated that the terrain was basaltic—like the most primitive lavas on Earth, such as the ones that well up from the mid-ocean ridges to form our ocean bottoms. Venera 8, however, had landed on a more complex rock, rich in alkalines and other elements which on Earth are present in various granites. Some of Basilevsky's colleagues at the Vernadsky have proposed that Venera 8 might have found something like syenite or diorite, which are similar to granite, except that they have no quartz. Some Western scientists have questioned whether there is granitic rock on Venus at all, and indeed the presence of the ingredients does not prove that they are in fact combined into granite. The possible presence of a granite-like rock, of course, has been used to further the suggestion that Venus, like Earth, has plate tectonics and even continents. But when Basilevsky looked at the flat terrain in the Venera 8 landing ellipse, he thought to himself: "There is certainly nothing that looks like a continent anywhere here, but there is this suspicious feature—this blini! It may be that Venera 8 landed on top of a pancake!" (Later, when Basilevsky was able to look at all the other Venera landing sites in the Magellan imagery, none of them had any trace of a pancake.)

I asked Basilevsky to tell me some of his recollections of the successful Venera landers. He remembered most vividly the first pair he had worked with, Veneras 9 and 10, the first to have cameras. He said, "My job, when I got to the control center in the Crimea a few days before touchdown, was to take the flight controllers' estimates of the trajectory and pinpoint as closely as possible the ellipse where Venera 9 would land. Was it in a flat area, or was it in a rough area? We still had time to change it. But the flight controllers' data indicated that it was close to the planned site, so we made no change. Then I awaited the touchdown. When the probe was in the atmosphere, there was a radio blackout, the way there is when spacecraft land on Earth. Then, after what seemed an eternity, we heard that it was O.K. I could hear the flight controllers reading out the temperature and pressures on the way down: 'One, two, three Earth atmospheres; ten, forty, eighty, ninety.' Then: 'We have contact!' We were on the surface, and everything was working.

"Everyone wanted to see the first pictures of Venus, a panoramic view. There were a lot of people, but no TV monitors, like you have all over J.P.L. There was just a single device that produced thermaprints. Because I wasn't on the imaging team, I was not allowed in the room with the thermaprint machine. There was a big crowd of people trying to get in, but the machine and the people on the science team were protected by security guards. The panoramic picture came out of the machine very slowly, band by band. When the first picture was halfway done, a friend of mine on the team spotted me and invited me in. I had to squeeze through the door without letting anyone else in. I saw a lot of plate-like flat rocks. A little later, a media person—we have them, too— asked me what I made of them. The chemical data wasn't in yet, but I said they reminded me of granite. My guess went out on the wires. But I was wrong. Two hours later, when the chemical data came in, it was basalt. The flat rocks may have been some sort of talus—Venera 9 landed on a steep slope, thirty-five degrees; any steeper, it would have tipped over. The spacecraft, after it landed, probably slid downhill on the talus, like a ski person." (Basilevsky is full of tales about the Veneras. On Venera 11 or 12, when the television port—analogous to a lens cap—failed to open, so the images were all black, one puzzled engineer remarked to another that the spacecraft might have sunk into a deep viscous fluid. Another, agreeing, responded, "We are definitely in the shit." They were.)

"My first impression of most of the Venera lander images was that they were all very boring," Basilevsky went on. "There was just some loose material and a few outcrops. Then I realized that at each site we imaged there is layering of the rocks—the layers are a millimeter to a few centimeters thick. It's most interesting. I cannot understand it. For me, it is not explained even now. If you look at the Magellan images, you see a lot of volcanic lavas. All the spacecraft landed in plains, and the plains are lava. But the Venera lander images do not look like lava. I feel a little uncomfortable because I don't see in the Magellan imagery this layered stuff! I don't see what can correspond to it." The only thing he could think of, he said, was a kind of volcanic rock called tuff, which is built up of volcanic ash; as the ash erupts and falls from the sky episodically, it can build up layers of rock; perhaps that was the explanation.

I asked Basilevsky, who had come closer to actually being on the surface of Venus than anyone else around, if he could describe from evidence provided by the landers what it would be like to stand on Venus—minus the temperature and pressure, of course, lest the viewer himself be squashed and fried into a pancake. The color of the surface, he said, is a dark reddish-brown (and not the fiery oranges of the J.P.L. images). The sun, though only 67 million miles away, is invisible, as it is on Earth in a heavily overcast day—an effect due not so much to the clouds, which are relatively thin, as to the thick atmosphere. The entire sky glows evenly, so that being at the bottom of the atmosphere of the bright Evening Star is like being inside a fluorescent light bulb. There are no shadows. There might be a breath of torrid wind—when Venera 13 touched down, a pile of dust landed on the base of the spacecraft. Over the next few hours, during successive panoramic sweeps of the camera, the dust gradually vanished—presumably blown away.

The only other atmospheric excitement might be an occasional flash of lightning—a phenomenon suggested by instruments aboard some of the Venera landers and also Pioneer–Venus, and confirmed by the Galileo spacecraft when it flew by Venus in

February 1990. On Earth, lightning is caused by strong updrafts carrying water droplets with opposite electrical charges, but Venus lacks both updrafts and water vapor. Its clouds, far from being thunderheads, are like a thick layer of fog—too quiescent, in the opinion of many atmospheric physicists, to generate much juice. On Earth, the other explanation for lightning is particles of dust tossed in the air by volcanic eruptions, which rub against each other and emit charges. Sulphur high in the atmosphere suggests that volcanism on Venus is a continuing phenomenon—though Magellan had seen no sign of active volcanism. The best chance of identifying a current eruption would come on later cycles of Magellan, when images of the same area taken at eight-month intervals could be compared.

THE CORONAE. Although coronae, bubbly domes most of which are between a hundred and a thousand kilometers across, were known from previous missions and ground-based radar, they also are specific to Venus. Unlike the much smaller pancakes, which are lava extruded onto the surface, coronae are caused by lava that pushes up the ground from underneath. Ellen R. Stofan, a young geologist at J.P.L. who was deputy project scientist under Saunders and who had recently got her doctorate from Brown, where she worked under Jim Head, had taken on the coronae as her special preserve. Though she was neither a principal nor a guest investigator, she was the leader of a group of scientists studying caronae, and the fact that this group was generally considered the one whose members worked best together was largely ascribed to her rather open and sunny ways. She, however, ascribed the group's success to the fact that most of its members were not spread all over the country but worked in the Los Angeles area, and they were reliable about answering their E-mail—the notes they sent electronically to each other's computers. (The group that reportedly worked least well together was the crater group, in part because of the rivalry between Schaber and Phillips. And the corona group subsequently lost some of its reputation for harmony when its members wrote three separate scientific papers because none of them would take a back seat to the others in the list of authors of a single paper.) It was being noised about that if a woman were ever added to the International Astronomers Union's committee of nine wise men picking names for Venus, Stofan would be an excellent candidate. A year and a half later, she was added to the committee, which lost its nickname.

Stofan showed me a couple of F-MIDRs containing coronae. "Here's one, Nightingale, which is 530 kilometers across," she said. "Here's another, Quetzalpetlatl, which is 900 kilometers across and a kilometer and a half high—it's more dome-like than Nightingale, which has a hole in it, like a big volcano. With the exception of Nightingale and another corona, Earhart, coronae are named for fertility goddesses. When I started getting interested in them, I got pregnant, and everyone wondered if there was a connection."

Brennan Klose stuck his head in the cubicle and asked Stofan if there was a Pomona Corona, and when she said there was (Pomona is a Roman goddess of tree fruits), he said his reason for asking was that he had seen a road sign that read: ← POMONA CORONA. → She explained that there is a California town named Pomona and one named Corona, quite near each other, and he had apparently been at an intersection between them.

In her opinion, and also in the opinion of Roger Phillips, coronae are the result of smaller plumes that originate at a possible discontinuity partway down the mantle. On Earth, there is one 670 kilometers below the surface, about a third of the way down the mantle, which reflects either some chemical change in the rocks or what is called a phase change, a geophysical term for what happens when increases in temperature and pressure reach a point where the crystals in the rock become denser and assume a more compact structure. At that point there would be a boundary akin to the one between the core and the mantle where instabilities could occur. Plumes that originate there would be smaller and weaker than the giant ones that originate at the core-mantle boundary and buckle up the far bigger regiones.

Coronae have much more definite shapes than regiones; they often look like huge blisters. Lava can ooze from the top. As they balloon outward, they create a pattern of fractures, the result of extensional pressures, radiating away from the center far into the surrounding plain. Like regiones, coronae have different stages, including the frankfurter stage with graben running down the center. "I hope to find lots of coronae, in different stages of development, and be able to plot their evolution," Stofan told me.

Increasingly, the scientists were feeling that all the circular manifestations of internal heat on the surface of Venus—the regiones, the coronae, and the calderas, and perhaps also the pancakes and some other recently discovered roundish features with long, leggy promontories which the scientists had dubbed ticks —were interrelated, with the deeper plumes being responsible for the bigger features and the shallower plumes for the smaller ones. Some of the scientists were wondering what might account for the different forms these features took. Later I found Ellen Stofan deep in conversation with Mark Bulmer, John Guest's graduate student from the University of London, who was studying Venus's calderas. They were interested in the fact that both calderas and coronae are circular, and both are apt to be surrounded by annular rings of trenches and ridges. The largest calderas and the smallest coronae are about a hundred kilometers across. "We're wondering if calderas are collapsed coronae," Stofan told me. She and Bulmer were trying to figure out what would cause a plume to make one or the other; it would have to do perhaps with the speed of the withdrawal of the underlying lava or the strength of the overlying material. In hot batter, sometimes there is a bubble that remains a bubble, and sometimes the bubble slumps to make a crater. With all these different types of bubbly features, the surface of Venus was beginning to look to some of the scientists a little like a slow-motion version of a hot spring—like, for example, one of those mud pots at Yellowstone National Park, where bubbles rise through the ooze, lingering on the surface, and in some cases burst to form cavities.

EACH of the LOSes, several flight controllers liked to point out, had coincided with the arrival from Washington of David Okerson, the headquarters liaison with the Magellan project. When I got to the Magellan office on Thursday, November 15, at 9:50, I could see the top of his head over a distant cubicle—he had flown in from Washington that morning.

There had been an LOS about three-quarters of an hour earlier, though the emergency was largely over now. The engineering section, as usual, was as quiet as a bank; a few people were popping in and out of their offices somewhat more rapidly than usual. As Cynthia Haynie, the sequencer, walked by, she filled me in. At 9:01 that morning, the spacecraft's signal was supposed to have been acquired after a mapping pass, preparatory to the transmission of the data—but the DSN antenna at Goldstone had failed to pick it up. The station had continued to listen over the X-band on the smaller, thirty-meter antenna for about half an hour before shifting to the large, seventy-meter antenna, which finally picked up the spacecraft's signal on the broader S-band at 9:43— the spacecraft's attitude turned out to be about a degree and a half off point, outside the X-band's half-degree beam width but well within the S-band's beam width of two degrees.

Haynie expected that before long she would be called upon to write the instructions to tweak it—a degree and a half was too far off for mapping and also for aiming the high-gain antenna so that transmission of data over the X-band with its half-degree beam width would reach Earth. She disappeared into her cubicle to start planning the tweak. Tony Spear walked by in the other direction. The X-band, he said, presumably was still transmitting data from the last mapping cycle. "The X-band thinks it's talking to Goldstone, but it's pumping all that stuff into space," he said. He had been at a meeting when the LOS had happened; when he was beeped, he thought it was an emergency at home—in the two quiet months since the last LOS, his fears of telephone messages had vanished.

The reason the LOS had lasted such a short time, Spear told me, was doubtless because of some important changes that Slonski and the other engineers had made in the heartbeat table, in light of their experiences with the earlier LOSes. When the delayed engineering data arrived several hours later, it turned out that there had indeed been a heartbeat loss at 8:07, three minutes after the mapping pass had started. This time, instead of swapping entire AACSes or going immediately into RAM safing, the spacecraft switched just one part of the AACS, the IODA, from A to B. The effect was the same—it caused a restart, the computer's version of the swift kick. Because the memory was not switched, the good memory with the correct guide-star coordinates remained on line. Again because the memories were not switched, the spacecraft did not go into RAM safing, but remained in RAM. Slonski and the other engineers had removed it from the ladder anyway, as they had been disenchanted with the performance of the RAM-safing program during the first LOS, when it had muffed a star cal and gone off on a lengthy wild-goose chase. As it happened, no safing program was needed, because the IODA swap did the trick, and after thirty seconds the spacecraft resumed normal operations. That was the true length of time of the LOS, and not the supposed thirty minutes, which was the length of time it took the DSN to shift antennas and bands. (Later, Slonski told me that he had serious questions about why the DSN had waited so long to do this.) Either way, compared to the thirteen hours and seventeen hours of the previous LOSes, which combined had caused a month's delay in operations, the third LOS was almost negligible; the engineers expected to be back mapping in a day or so. The damage control was entirely due to the changes in the heartbeat table. Still, the third LOS was unpleasantly similar in kind to the earlier ones, and the engineers were no nearer understanding the cause of the problem.

I followed Spear into the conference room where the engineers normally met twice daily, at 10 a.m. and 5 p.m. The morning meeting—my first taste of NASA crisis management—was about to begin. As long as the engineers were unable to send commands reliably to the spacecraft, the situation was still critical, and so the atmosphere in the room was hot and heavy—truly Venusian. Three tables arranged in a horseshoe filled half the room. James Scott, the mission director, sat at the head, with a score of other engineers arranged around him, including Kenneth Ledbetter, the deputy manager from Martin Marietta, Douglas Griffith, the deputy mission director, and Slonski. They were joined shortly by Haynie. I sat in the first of two rows of chairs at the back, near Tony Spear, who was talking with an engineer sitting be-

tween us, William Johnson, the chief engineer for
the radar. Spear was worried about how they were
going to correct the spacecraft's attitude. The one-
and-a-half-degree pointing error represented the
amount by which the attitude had drifted during the
nine seconds' delay before the computer was reset.
The gyros were oblivious to the change—they thought
they were pointing to the same spot in the sky as they
were before the heartbeat failure. The gyros had
swapped from A to B along with the IODAs, and the
backup gyro was even more out of line than the
primary.

"We're due for a star cal, and that will screw
everything up," Spear said. It would be like trying to
navigate a boat that was off course, using a compass
that did not point north.

Johnson said, "If the spacecraft is a degree and a
half off point, the star scanner won't pick up the star,
and it won't have a detrimental effect—there will be
no update of the gyro, and the spacecraft will just
maintain its present attitude."

Spear did not look any happier.

At the front of the room was the squawk box
connected by a telephone line to the other squawk
box in Denver. The two groups of engineers were
trying to agree on what the status of the spacecraft
was. The J.P.L. squawk box, a beat-up gray plastic
affair which looked as if it had been bought years
earlier at a five-and-ten-cent store, was squawking
more than talking.

Ledbetter, who was trying to check the spacecraft's
systems with Denver, was saying, "The power system
is nominal."

The Denver engineers asked him to repeat that three times before they got it.

Then a woman in Denver, Julie Webster, said, or tried to say, "The propulsion system is normal. We haven't used any thrusters."

Everything Webster said, Griffith asked her to repeat. Finally Scott broke in and told them to try to fix the communications from their end while they did the same here. A J.P.L. engineer took the squawk box down from its shelf, shook it, and put it back— the equivalent of the swift kick just administered in space by the IODA swap; there was the sound of something similar being done in Denver. The transmission was no better. Transmission with the spacecraft out around Venus was far clearer; at least, it was when the spacecraft was pointed correctly. Finally Scott told Denver to hang up and dial again—a new circuit might be better. It was, marginally.

Ledbetter asked Denver, "Is anyone there working on sequencing for turning off the tape recorder?" Johnson explained quietly to me that the tape recorder used for the mapping data also contained on its engineering channel data that had been recorded before the LOS, and if they could retrieve it, the data might reveal the underlying cause of the problem. The tape-recorded information was different from the recorded engineering data—the memories' log of events—which they had gotten after each LOS, in that it was more narrative in nature, and also included events prior to the start of the LOS which might give a clue to how the trouble had started. The engineers had gotten back some taped data from just before the first LOS—it had not indicated that anything whatsoever was wrong before the heartbeat was lost; it was important to try again. However, if the recorder wasn't turned off, the data would be erased and overwritten.

"It's essential to get that engineering data back," Ledbetter said.

Denver said they would look into it.

"We'll get Cynthia Haynie cranked up at this end," he said. Haynie, who was sitting on the left side of the horseshoe, rose and left the room.

The squawk box suffered another bout of static, during which Denver was LOS. While it lasted, the J.P.L. engineers looked resignedly at the table in front of them. When Denver's walkabout ended, the engineers talked about ways of tweaking the unaligned gyros, the first step in correcting the spacecraft's attitude. As Johnson had told Spear, and as Webster in Denver was now announcing to the room, as long as the spacecraft's gyro was inaccurate, the spacecraft would be unable to point its star scanners near enough to the guide star for the star scanner to see it—or, as the engineers said, "accept it." The window of acceptability of the scanners had been set very low, at .07 degree, to reduce the possibility of error, such as accepting the wrong star or a piece of brightly shining Astroquartz as the guide star. The engineers proposed now that a command be sent to open the window almost ten times wider, to .9 degree, to allow the scanner to accept the star. They also decided to send commands to switch back to the more accurate primary gyros and to read out memory A.

The engineers talked about the possibility, if the gyros couldn't be aligned, of orienting the spacecraft by commanding it directly to point toward Earth, using for guidance the amount by which the reception of the spacecraft's radio signal was low. It was the weakness of the signal that had told them, to begin with, how far the spacecraft was off point, and Julie Webster in Denver thought they could also use it to ascertain in which direction the spacecraft was off, and then they could command the spacecraft to turn in that direction. This method, which they called the manual method of changing the spacecraft's attitude, was like the dead reckoning used aboard a ship when its compass is out of whack.

Slonski started to recommend a change in the heartbeat table. However, he was cut off by a booming voice from the ceiling, where there was a loudspeaker that was part of J.P.L.'s center-wide intercom system—a woman with an authoritative voice was reading a lengthy announcement about the Thanksgiving holiday, a week away. Slonski drew his index finger slowly across the front of his neck. As the voice droned on, Tony Spear suggested that they cut the wire. The Denver engineers learned more than they wanted to know about Thanksgiving in Pasadena.

At 10:25, after a minute's LOS, conversation resumed. "Let's break up," Ledbetter said, scraping back his chair. "I think we've all got enough on our plate to start working." The room emptied.

At 10:30, I looked into Cynthia Haynie's cubicle. She was too busy to talk—she was hard at work on the command sequences for turning off the tape recorders.

At 10:40, the DSN sent the commands over the S-band to turn off both the spacecraft's tape recorders.

....

They were trying to get in through the low-gain antenna, the small dish standing in the center of the big dish, which is capable of receiving S-band signals from half the sky. Even so, they didn't get in. The information on the tape recorders was lost.

At 1:52, the DSN sent a second set of commands over the S-band to increase the window for the star scanners, to switch to the primary gyros, and to tweak the spacecraft's attitude in the direct, manual manner. These commands did not get in either.

At 1:35, I peered into a small conference room, where Ledbetter, Griffith, Slonski, and several others were holding an animated conference. They were talking about making changes in a tape repeating the same commands that had failed to get in at 1:52, to be sent up later that afternoon, and maybe using the X-band. Slonski was arguing for making the commands as brief as possible. As he had done at the time of the second LOS, he wanted to make sure they gave it their shortest, hardest shot.

At 3:13, the commands were sent, over the X-band this time. It was a lucky shot. Because of the X-band's narrow beam, it could be concentrated with great intensity; and because the DSN was using its biggest antenna and its highest power, the transmission got into the spacecraft's high-gain antenna through what engineers call a side lobe. (Just as antennas can leak signals from directions other than the intended one, they can sometimes receive them in this way, too.) The reason the earlier commands sent on the S-band had failed to get in, the engineers now realized, was that Venus was still so close to the sun as seen from Earth, following the period of superior conjunction,

....

that the sun's corona interfered with the S-band's signal, which was more susceptible to this effect than the X-band. "It was like attempting to communicate with the spacecraft through the flame of a rocket plume," Spear said later. Earth had been able to detect the spacecraft's S-band carrier signal despite the sun, because the DSN's powers of reception are so much greater than the spacecraft's.

When I passed Okerson in a corridor, I asked him whether he felt getting the commands in had been a matter of luck. He didn't think so. "It's a matter of trying and trying and trying again," he said. "That's what we do here."

Though the star scanners' window had been widened, the spacecraft still failed one star calibration after another. At a little after four, an engineer in the conference room was chatting with Denver over the squawk box. "How many star cals can we miss before we go into ROM safing?" a Denver engineer asked.

The J.P.L. engineer thought it could miss a total of six—after that, the spacecraft would go into ROM and start coning, unless that safety provision was overridden by the ground.

I sought out Okerson again and asked why the star cals weren't succeeding despite the wider window. "We opened the window of acceptability for the error in the star's position, but we did not open the time window for how long it takes for the star to pass across the two slits as the spacecraft rotates," he said. This factor is as important to acceptability as the position. "We probably should have widened that

. . . .

window along with the other, but you don't want to open all the windows, or some dirt will fly in. It will increase the likelihood of an erroneous star cal." He expected the time requirement would be met on the next cal, assuming that the command to tweak the spacecraft's attitude manually, which had been sent again, would work.

At 6:45, I ran into Okerson again, and he told me that the manual tweak had been mostly successful—the spacecraft was only half a degree off point, close enough for mapping Venus. All hands, he said, were pleased with the results of the changes in the heartbeat table, and in safing, which had been carried out after the previous LOS.

At a little after 7:30, the engineers gathered again in the conference room to discuss the safest way to moor the spacecraft for the night, now that it was nearly back in harbor. The major issue was whether or not to return the spacecraft to the top of the heartbeat table, in which case it would be restored to life with another IODA swap, or leave it where it was, one step down the table, for the night—in which case another heartbeat failure would send it into memory B and very possibly enter ROM safing.

At 7:55, a phone rang in the conference room. "It looks like we got a good star cal," the engineer who answered it said, hanging up.

"Right, we got a good star cal," Scott, the flight director, said. "Denver, we got a good star cal."

"Right. Let's go on with our discussion," Slonski said. I was surprised at the matter-of-fact way in which the engineers accepted the news that their

. . . .

earlier efforts had been crowned with success. Flight engineers have learned to concentrate on the present problem, to the exclusion of all else.

Before they broke up, they decided to leave the spacecraft as it was for the night. "Jesus, I don't like that!" Okerson, who had been vehement on the other side of the argument, said as he left. "So we'll spend tonight at risk of going into ROM safing—it'll take a week to get back to mapping if we have another failure." The night, however, passed without incident.

WHEN I returned to J.P.L. in the middle of February, the only change in the science room was a big punching-bag plastic balloon of Gumby, which the scientists would hit on their way to and from the bathrooms just beyond—"to take out our frustrations about Venus," Saunders told me. It had been donated for this purpose by a member of the science support staff. Otherwise, the room was quiet, and remained so most of the week I was there. There weren't that many scientists around, and those who were were working by themselves. I missed some of the excitement of the first week—and of earlier missions, like the Voyager flybys, which I had been present for. Flybys, especially, pack a lot of action into a very short time, but even the Mariner and Viking missions to Mars, which like Magellan were long-term orbital ones, had managed to sustain considerable suspense. Perhaps it had to do with the thick carpet on the floor—for earlier missions, there had been linoleum or composition tile, which made the sounds reverberate. Part of it, Head told me when I saw him, was the cubicles, whose walls went only halfway to the ceiling; this inhibited the sort of loud talk of Voyager or Viking—or Galileo. "Last December, when Galileo flew by Earth and the moon for a gravity assist on its way to Jupiter, I was in charge of Galileo's imagery of the moon," he said. "It involved a lot of teamwork, of people working hard together. It was much more like a Voyager encounter. You compare the half-day encounter of Galileo with a half-day of Magellan at Venus, and the difference is quite remarkable."

Part of the lower intensity of Magellan, he felt, had to do with its low cost, which did not provide for long-term residence in Pasadena by the scientists. The scientists came to J.P.L. for a week every month or two, picked up the F-MIDRs and F-BIDRs (and, later, the compact disks), and went home. In some cases, they didn't come at all and the data were mailed to them. Most of the time, the only scientists at J.P.L. were Saunders, Solomon, Stofan, and Head. Both Head and Saunders, whom I talked to a little later, missed having the opportunity to bang around ideas with their colleagues. With so many people absent so much of the time, some members thought the team lacked cohesion—even collegiality. Another person who missed the old intensity was Jerry Schaber. "The spirit is different from other missions, when we worked more as a team," he said. "On Magellan, people work more on their own. They grab what they want and disappear." Schaber was in a bad mood, because he hadn't been to J.P.L. for a month, and he had a hundred orbits of crater counts to do.

Another factor that made Magellan quieter, Head

thought, was that Magellan was not the first radar mission to Venus, but the fourth. "That means that before Magellan arrived, the level of knowledge about Venus was quite sophisticated," he said. "Instead of asking first-order questions, which tend to be the most immediately galvanizing, we are asking second- or third-order questions. We're drawing people in from different fields, like geophysics, as opposed to having just a bunch of geologists sitting around in a back room. Overall, Magellan may be less intense, but in many ways it's more rewarding."

Magellan had completed mapping western Aphrodite and was beginning on eastern Aphrodite, the scorpion's long tail that terminated in Maat Mons, the sting. By now it had mapped seventy percent of the planet—a distance equivalent to that between Pasadena and Peking, by way of the Atlantic Ocean. The processing of the images, however, was about a month behind because of the slipped-bit problem and other difficulties. The F-BIDRs and F-MIDRs for the Ovda region, where Jim Head had thought he had seen, on Pioneer–Venus and Arecibo imagery, signs of rifting and crustal spreading, had just been completed. "I don't see any of this in the data!" Steve Saunders told me almost gleefully, when I saw him —he was not a believer in Head's theory. He showed me an F-MIDR of Ovda, which he had not had much time to study. It was a complex mishmash of terrains. On a later occasion, when Saunders had had more time to study Ovda, he pointed out some of its complexities. The northern edge dropped off to the plains in just a couple of kilometers. As he moved his finger farther south, there came a region of mesas which got brighter as they got higher. The most mountainous parts had the highly reflective white material associated with high altitudes, such as Sif and Gula Mons and Maxwell. Then, as he moved his finger farther south still, there was a lot of tessera terrain. In the southernmost zone there was a network of fine filamentary fractures. He found a crater on a rille. The southern margin of Ovda, where it dropped down to a plain, was creased with troughs and ridges—which he thought were the latest tectonic developments in the area. The whole circular region, he thought, looked as complex as the surface of the brain. He had no idea how to explain what he saw, but he was quite sure he was not looking at a spreading center.

Jerry Schaber, when I saw him a little later, told me that the latest Magellan data showed that there was no increase in crater rates, and hence in age, moving northward from the equatorial highlands, as had been suggested by the Venera 15 and 16 data, which had shown a decreasing rate down to the thirtieth parallel. Head found this evidence discouraging to his theory. Even in the absence of plate tectonics, he told me, if there was spreading from the equatorial highlands, the terrain should still get older as you moved away from the spreading center. Moreover, the figure for the age of some of the oldest plains of Venus now appeared to be as much as 800 million—far older than the plates on Earth, where the oldest age of any part of the ocean bottom is 200 million years. Sean Solomon told me, "I think the plate tectonics model is dead. We've been to the places where Head said the evidence for spreading should have been, and we don't find it. I don't think there is reason to expect spreading centers in eastern Aphrodite, where the long chasm is. High topography wants to become low topography, and as it falls down it pulls apart and chasms result."

Basilevsky, who at Brown–Vernadsky symposiums and elsewhere over the years had argued with Head against the idea of spreading, was pleased with the Aphrodite data. "Even if the evidence was covered over with mud or flows, the ridges and the rifts and the transform faults are so big you can't hide them if they're there," he told me. "My doubts about spreading on Venus and my arguments with Head have some history. Pioneer–Venus, though it imaged the whole planet, did not have enough resolution to distinguish delicate tectonic features, such as tesserae. At the time of Venera 15 and 16, which could make this distinction, we thought it would be useful to know about tesserae outside the northern quarter of the planet. So we compared the tesserae there with the same places on Pioneer–Venus—on radar, they give a specific combination of high surface roughness and low reflectivity. We taught our computer to look for these signatures on the Venera data, and then we asked the computer to search for these signatures on the Pioneer–Venus data for the rest of the planet. And the computer predicted that both Alpha Regio and Aphrodite would turn out to be tesserae. When the Magellan imagery of Alpha Regio came in last fall, our prediction that it was tessera was proved right. And now that we see Aphrodite on Magellan, I am satisfied that we are right again. Tessera is a garbage-bucket definition—it simply means grooves and ridges caused by deformation intersecting in two or more directions, usually on a plateau. But the garbage-bucket does not include the possibility of a spreading center."

Jim Head was by no means ready to throw in the towel—though he was too good a scientist to get bogged down in the question. When I asked him about the Aphrodite images, he said, "For me and most of us, it's a sensory overload! I just saw it this morning for the first time. I'm really at a loss for words. It's just beautiful. I feel like Charlie Duke, the Apollo 16 astronaut, when he got to the Fra Mauro highlands on the moon. All he could say was 'Fantastic!' He said it twenty times. Aphrodite is awesome in its detail. It looks like a continent. It makes me think of hot spots—and it also makes me think of spreading. You see pieces of evidence for all these ideas, but no evidence that pushes you in the direction of one over another. And, as Steve Saunders said recently, it may be none of the above." He said the terrain was far more complex than anyone had imagined, but still he thought he could tease out evidence for spreading, though it was hard to make out, as there were other things going on as well; what he had thought might be transform faults on the Pioneer–Venus imagery, on the Magellan imagery was overlaid with other patterns. "As you know, we've proposed a hot spot there, like Iceland—and Iceland is on a rift, with spreading, so maybe we're dealing with something like that." He was still waiting for the images of eastern Aphrodite, where what he thought would be the best examples of a rift with a central rise and transform faults would be.

BOTH Saunders and Solomon were increasingly in agreement with the plume theory. Solomon, a tall, spiky stalagmite of a man, got his doctorate in geophysics from M.I.T. in 1971; he has remained there ever since. Solomon, as chairman of the geology and geophysics task team, and Head, who was vice chairman, effectively administered the scientific work of the science team, while Saunders, the project scientist, was more concerned with the team's relations with the engineers and J.P.L. Solomon and Head shared an apartment when they were in Pasadena.

"The first year I was a graduate student, plate tectonics caught on—and then came the lunar landings and the unmanned missions to the planets," Solomon told me. "I became enthralled with these two parallel threads—I've been involved in both ever since." Solomon had a more detached, scholarly air than most of his colleagues, and I frequently relied on him as a sort of indicator for what the scientists were thinking. Despite his roots in plate tectonics, Solomon was so sure of the importance of the plume theory for Venus that he was shifting his attention to one of its ramifications—what happens to large-scale structures like the Ishtar highlands, with Lakshmi plateau and Maxwell Mons, when the plume that is holding it up dissipates. "Will it then be transformed into the sort of feature we see elsewhere?" Solomon asked. "Should we find old Ishtars in other places? The likeliest candidate is Ovda Regio in Aphrodite. It's a candidate for an old hot spot." (Though on many Venus maps Ishtar looks far bigger than Ovda, this is an illusion due to the projection, just as on some terrestrial maps

Greenland looks as big as Europe; Ovda in fact is bigger in area than Ishtar.) Solomon was beginning to think that when whatever was holding up Maxwell dissipated, the mountain (because of the heat of the rocks and the unusual steepness of Maxwell's flanks, which argued that it was very young) would spread out laterally, its sides cracking with graben, and crumble, losing altitude rapidly—within a few millions, or tens of millions, of years.

He, Phillips, and others were also getting interested in sorting out which topographical features were caused by upwellings and which were caused by a related phenomenon, downwellings. Downwellings are the other side of the coin from upwellings. Just as plumes rise up from the hot thermal boundary layer where the core and the mantle meet, the cooler boundary layer just beneath the lithosphere yields viscous rock that sinks into the interior. These downwellings take place sometimes in great descending sheets or sometimes in tighter cylinders called antiplumes.

As with upwellings, downwellings were first thought of for Earth—only, on Earth there is very little evidence of at least the cylindrical type of downwellings visible on the surface. Upwellings are far more noticeable because they intrude lava into the crust, causing great domes, seamounts, and chains of volcanoes like the Hawaiian Islands. The subducting plates on Earth can be thought of as sheet-like downwellings where the cold thermal layer extends to the surface in the ocean basins; and of course the visible evidence on Earth for subduction—ocean trenches, volcanic arcs, island arcs, mountain ranges—is quite striking. But with respect to the cylindrical type of downwellings, their effect on Earth is muted by the soft rocks of the asthenosphere; the evidence for them comes from the arcane instruments of geophysics—gravity measurements, seismic measurements, and measurements from strain gauges.

The idea that cylinder downwellings might have a more obvious effect on the surface of Venus than of Earth, and that sheet-type downwellings might occur there even in the absence of plate tectonics, was suggested in the middle nineteen-eighties by Duane L. Bindschadler, who was then a graduate student at Brown and is now a geophysicist at the University of California at Los Angeles. (Many graduate students, including Bindschadler and Klose, with his prediction of the pancakes, contributed some of the more original and insightful thinking to the study of Venus.) If Venus had no asthenosphere, Bindschadler reasoned, then the effects of convection, and in particular of downwelling, should be far more noticeable than they are on Earth. "When we tried to make a mathematical model for what the result would be, and ran it through a computer, we found to our surprise that a downwelling could cause a topographical rise," Bindschadler told me, when I saw him later in Pasadena. At first, there would be a depression, as the lithosphere was pulled downward. But afterward, as the downwelling continued and became stronger, the crust on all sides would be pulled inward and (instead of being sucked downward) would pile up and up, to create high topography. In later stages, the inward-moving lithosphere would press harder and harder toward the rising edifice, resulting in ridges and other signs of compression. (Downwellings and upwellings work the surface in opposite ways. Plumes at first cause topographical highs, but later these highs tend to fall over and collapse, because the upwelling material spreads outward along the bottom of the lithosphere, exerting a more extensional force on it, pulling it apart, with lava forming new crust. Indeed, on Earth many rifts are caused by upwellings.) Bindschadler, a lanky, uncompressed-looking scientist in his late twenties with a ponytail, told me that he believed the more amorphous, sheet-like downwellings caused the tessera plateaus and the cylinders caused the more circular features, such as some of the regiones—Alpha, for example—and maybe even the circular areas of western Aphrodite, such as Ovda and Thetis. Bindschadler said he had had something of an uphill battle maintaining these ideas at Brown, where the spreading theorists regarded them as heretical. If he was right, though, he was pointing out one of the most significant differences between Venus and Earth.

Bindschadler feels that downwellings would have a greater effect on the surface of Venus than would upwellings, because their force would be more concentrated there—in particular, the small cylinders breaking away downward from the crust-mantle boundary, where they would have a greater effect than the more diffuse plumes with their broad mushroom heads. Bindschadler believes that Ishtar and especially Maxwell are the result of a late-stage downwelling, with Maxwell having been thrown up and at present being held up by the great compressive forces—though Roger Phillips told me that he still felt a young upwelling fitted the facts better. (Solomon sides with Bindschadler.) Some scientists thought that coronae might be caused by downwellings instead of upwellings, though Ellen Stofan disagreed. Indeed, there is considerable disagreement among the Magellan scientists about which features are caused by upwellings and which by downwellings. Is a particular depression the result of an early-stage downwelling or a late-stage upwelling? Is a particular rise the result of a late-state downwelling or an early-stage upwelling? There is, however, near-total agreement that both forces are at work.

The only way to tell the two apart is by gravitational readings done by the careful tracking of the spacecraft as it passes over the various features to determine their gravity and hence their depths of compensation. Such gravity measurements require close analysis of the slight change in the spacecraft's carrier signal known as the Doppler effect as it speeds up and slows down over areas of greater or less mass. The shallower the depth of compensation, the lower the gravity of a feature of a given relief, because the area of lesser density is nearer the surface. With a feature caused by an upwelling, the depth of compensation is deeper, and hence the gravity—and the spacecraft's speed—greater than with a similar feature caused by a downwelling. The reason why Phillips believes that Ovda and Thetis are caused by upwellings is that their gravity—as measured rather crudely by Pioneer–Venus in the early nineteen-eighties—seems to be on the high side. (The Veneras produced no gravity measurements because they were oriented and stabilized with thrusters, instead of using momentum wheels like Magellan or simply being spun on their long axis like Pioneer–Venus, and the constant little bursts from the Veneras' thrusters imparted a velocity which overrode the effects of gravity. In addition, NASA's DSN is more accurate at measuring subtle changes in a spacecraft's speed than its Soviet counterpart.) For the most precise gravity measurements as well as for the most accurate global coverage, a circular orbit is needed, but the previous spacecraft were in elliptical orbits. The thinking was that Magellan would go into a circular orbit in a couple of years, after the end of its fourth cycle. Normally, this is done by firing a large rocket, but Magellan did not have one for this purpose, and its thrusters were not powerful enough for more than slight orbital changes. Instead, the plan was to use a relatively untried maneuver called aerobraking, which meant dipping the spacecraft repeatedly into Venus's atmosphere to slow it and lower its apoapsis.

(The spacecraft had not been designed for this maneuver, either, and consequently lacked the necessary shielding against frictional heat, but engineers who were working on the problem thought they could manage it. Aerobraking to circularize an orbit had never been tried before in either a shielded or an unshielded spacecraft, and there is a considerable risk of overheating. The plan was to use the big high-gain antenna as a shield, entering the atmosphere antenna-first, with the spacecraft snuggled behind it, like a person holding an umbrella in front of him during a driving rain. The operation would be conducted a little bit at a time over a long period—about seven hundred orbits over a hundred and ten days.

First the spacecraft's thrusters would fire backward at apoapsis to slow the spacecraft and lower periapsis from its height, at the end of the fourth cycle, of between 180 and 200 kilometers above the surface to approximately 142 kilometers above the planet—where the atmosphere begins. After that, each dip into the atmosphere would slow the spacecraft so that its apoapsis, 8,458 kilometers above the planet to start with, would spiral downward—Magellan's elliptical orbit would be wound down and down toward the planet, a little like a watch spring being wound tighter and tighter. In all its dipping, the spacecraft's temperature should never exceed 120° Celsius. When apoapsis was at the right altitude, about 250 kilometers above the surface, thrusters again would be fired there, in a forward direction this time, to speed up the spacecraft and raise periapsis to the same height as apoapsis. The orbit would now be a circle.)

Whether the gravity measurements proved that individual circular features were caused by upwelling or by downwelling, Venus with all its convection was beginning to sound more and more like a pot of boiling water or like the mud pots at Yellowstone—though pots of either sort are not properly spherical. Nearer the mark, perhaps, would be a spherical ball of mud such as an astronaut might levitate in a spacecraft—astronauts love to play with spheres of water or orange juice in space, where the liquids form perfect globes just like tiny planets. In time, the outside of the mudball would harden, forming a cool, solid crust. Imagine that in the center of the mud ball there is a hot electrical filament, like a planet's core, which the astronaut turned on. Instant Venus.

ONE saying going the rounds of the science team was that Venus was a volcanist's delight but a structuralist's nightmare. I asked Solomon, whose interest in tectonics makes him a structuralist, what this meant from his point of view. He gave me a different version of Head's view of Venus as a great big road cut. "In the images, I am struck by the difficulty we're having seeing through the sequence of events," he said. "Unlike Earth, where there is erosion, the entire tectonic and structural history of Venus's surface, at least as far back as 800 million years [the new estimate for the oldest parts of Venus], is preserved, and though this is an advantage, it makes problems. Venus is like a tape recorder that doesn't erase, and successive episodes get recorded on top of each other." I had recently been to the La Brea Tar Pits, which are a few miles away from J.P.L., where thousands of paleolithic animals—lemurs, saber-tooth tigers, mammoths—had sunk to oily deaths, and I had watched through a window some paleontologists chipping away at the masses of beautifully preserved bones, many of them all mixed up, that today are mined from the dried tar beds. (Indeed, the surface of the tar pits look a little like a speeded-up version of the surface of Venus, with little bubbles roiling the surface.) I asked Solomon if his job on Venus was like the La Brea paleontologists', and he said it was. It was this sort of tedious teasing apart of the structure of Ovda that he would have to do in order to see if it might once have been an Ishtar.

On a smaller scale, he was still fascinated by the systems of grooves and ridges that could look as chaotic as bundles of spaghetti or as orderly as the ripple marks on the ocean bottom. These structures, which had struck the scientists from the first week, were signs of tectonic stresses—of the pulling-apart of the lithosphere here and the pushing-together of the lithosphere there. In the absence of an asthenosphere, most of the scientists agreed, the powerful convective forces in the mantle are coupled directly to the lithosphere, and may cause it to move this way or that—the source of all the signs of horizontal motion, of compression and extension of the crust. The upwellings and downwellings in the mantle not only heaped up the lithosphere here and there but also pulled it apart, forming faults and mountain belts. As time passed and the upwellings and downwellings evolved or died out and new ones appeared, the stresses would come in different places and different directions, causing a terrain that was ridged or grooved in one direction to be ridged or grooved in another—the tessera.

The grooves and ridges that were organized into long, wiggly forkfuls of spaghetti—ridge belts, as Solomon called them—interested him particularly. They were evidence suggesting that the lithosphere had different layers—that it was what he called a jelly sandwich, with a more rigid top layer of crust, a squashier lower crustal layer as temperatures mounted, and then a more rigid layer at the top of the mantle. A geophysicist with the Goddard Space Flight Center in Maryland, Maria Zuber, had shown that as the lithosphere was compressed horizontally, the lower layer crinkled in larger undulations to form the bigger features, the belts, and the upper layer crinkled more delicately to form smaller grooves and ridges which ran through the belts.

The lithosphere, in fact, was in constant, if infinitely slow, turmoil. One geophysicist on the Magellan science team, William L. Kaula of the University of California at Los Angeles, called the process scum tectonics, and indeed the wrinkles and plains of Venus's surface resemble what the scum on a cup of boiled cocoa would look like if you moved it cautiously about with a spoon. (When I told Basilevsky a little later about the scum model, he asked with a puzzled frown, "What is this word 'scum'?"; and when I explained, he nodded, and said, "Oh. It is *pienka*." He added it to *blini* as a new geological term. He preferred a boiling-water model with a thick coating of oil on top to the cocoa and *pienka*.) Scum, oil, or lithosphere—if the lithosphere were broken, there would be rifts, and Venus had its share of them, many in the equatorial highlands. Unlike rifts on Earth, which through crustal spreading can broaden to form an Atlantic Ocean, the rifts on Venus had not split apart more than a few tens of kilometers. In the middle of April, when the spacecraft was over Beta Regio, it imaged a rift that had run through a crater and widened, so that one piece of the crater was ten kilometers west of the rest of it. Solomon referred to the rift as a "failed Atlantic Ocean."

The idea of the lithosphere scumming around appealed to everyone. A month later, at the planetary-science conference in Houston, Solomon said, in a talk he delivered, that there was evidence on Venus for horizontal movement of the crust from a few hundred meters to tens of kilometers to hundreds of kilometers—but none in the thousands of kilometers, as is the case with plate tectonics on Earth. He believed that the largest-scale motions on Venus—the ones in the hundred-kilometer range—were due to internal convection. Phillips told me that he was in full accord with this proposition.

I ASKED John Guest whether he agreed that Venus was a volcanist's delight, and he nodded enthusiastically. He had found a great variety of volcanoes on Venus, running the gamut of the shapes found on Earth, though many of them were much bigger. There were more and more signs of pyroclastics. What he found particularly interesting, though, was the extreme length of the lava flows—far longer than any on Earth, because the high surface temperature on Venus kept the lava liquid longer, delaying the solidifying of the front of the flow, the first part to cool, and preventing the flow front from damming up the lava behind it until the flow had traveled much farther. He showed me one caldera with a big delta of lava extending three hundred kilometers from it. Most of these myriad signs of volcanism were caused by heat in the upper part of the mantle, as it is on Earth; however, clusters of volcanoes, of which there were several, might have had a common source in a plume.

Guest was sitting at a drafting table helping three or four graduate students who were mapping flows and craters and other features on Venus. Mapping is the grunt work of planetary exploration—and, like most grunt work, it is the foundation on which much else is built. (It, like crater counting, appeals to geologists, who regard maps as what they call their basic data set, but mapping bores many geophysicists out of their skulls.) The maps were contiguous and meant to be put together, but the students were having trouble agreeing on the symbols they would use to identify impact craters, volcanoes, coronae, and ticks, and they were not all in step. "I hope we'll get a consensus," Mark Bulmer said. "As we talk day after day, and compare what we're doing, we get closer together. The problem is not just deciding on the meaning of the symbols, but deciding what a feature is in the first place. For example, is this a pancake in a degraded state, or is it a crater? It's hard to tell. The idea of mapping by hand instead of relying on the SAR images is to provide information, and that means you have to make decisions, as you go along, about what is what. And the rules of photo mapping no longer apply with radar. What we think is a morphological boundary might instead be a textural boundary. So we have to be very careful to see what things are." Bulmer said he planned to do his thesis on the relationship between calderas and coronae and the different circumstances that led to the formation of one instead of the other (the subject on which he had been conferring with Ellen Stofan); and careful mapping of as many of them as possible was the first step.

Next to him was Kari Magee-Roberts, a student of Jim Head's, who was mapping an area called Lavinia, which contained a feature that might be a rift. She planned to research it for her thesis. The apparent rift ran east and west on top of a ridge, and on one slope of the ridge was a volcanic vent. From it extended a series of flows—collectively, they were called Mylitta Fluctus—that pooled in the plain below; from the gigantic puddle extended individual flows like fingers, some of which were 800 kilometers long. Lava had poured from the vent on many different occasions, and Magee-Roberts was trying to do a stratigraphy of the whole area, seeing if she could untangle which flows overlay which other flows and hence were younger. So far, she had found four separate ages, and thought she could make out a fifth.

On the other side of the table, a geologist, Henry J. Moore, a soil-mechanics expert with the U.S.G.S. at Menlo Park, was working on a map with another of Guest's students, Michael Lancaster. (Menlo Park, a town on the peninsula south of San Francisco, is the original home of the astrogeology branch where a few stubborn astrogeologists who refused to move to the wilds of Flagstaff still hold out.) "We need more symbols to describe craters under twenty kilometers," Moore said. "The trouble is that, on maps below a certain size, symbols like diamonds and circles and everything else look alike."

Lancaster went around the table to see how Kari Magee-Roberts was doing with Mylitta Fluctus. "Everyone says Olympus Mons on Mars is the biggest volcano in the solar system," he said, inspecting the big glob of lava flows. "It isn't. Venus is. The entire planet is one big volcano."

. . .

THOUGH the engineers' quarters always seemed to have a bank-like hush, it was a bank that seemed to feel constantly on the verge of bankruptcy because its main asset might disappear at any moment without notice. As always, Slonski and Okerson had anxious looks. By early February, all four channels of tape recorder A had failed and the spacecraft relied solely on tape recorder B—a situation NASA engineers call single string. If the single string broke, there would be no string and no more data from the radar. The lifetime of a recorder such as this is seven thousand recordings and playbacks, and Tony Spear, with Pettengill and Saunders, was trying to figure out, if they were unable to fix recorder A, whether they would use up recorder B on the next two eight-month mapping cycles, or would be able to stretch it out over a longer period. One worry was that the problem in recorder A might somehow spread to recorder B —but there was some confidence that it wouldn't, for this particular type of recorder had been used successfully on sixty previous missions, mostly military ones. It was a special type, used for handling vast quantities of data, and hence was referred to in the trade as the Gigabyte Recorder. It was also used aboard Galileo and the Hubble space telescope. "As you can imagine, the Galileo and Hubble people are looking anxiously over our shoulder," Okerson told me. Slonski had had a few days' panic, for just after recorder A had been turned off, recorder B had seemed to be starting to evince the same symptoms. However, the symptoms were present only when a particular antenna at Goldstone was receiving the data, and the problem turned out to lie there and not with recorder B. Meanwhile, work was going on at Odetics, the company that had made both recorders, to see if they could figure out the trouble on recorder A and fix it. The company's engineers had suspected the problem might have to do with the springs on the two reels that put tension on the tape, but it wasn't that. Then they thought the problem was faulty erasing, but that turned out not to be the case, either. Then in June they tracked the problem down to an insufficient current in the recording head. Later in the summer, they traced the insufficient current to a short circuit in one of four capacitors which store an electric charge. There was nothing they could do to fix it.

In addition to the tape recorders, the spacecraft was already single-string with its memories and perhaps also its processors. About three weeks later, the spacecraft developed a loud whistling caused by a malfunction somewhere inside transponder B, one of the two redundant communications systems which include not only the transmitters but the amplifiers, the modulators that put the data onto the subcarrier, and other components. The spacecraft was now single-string on transponders, too, until or unless the engineers could repair the whistle. The spacecraft was beginning to resemble an old jalopy in ways other than its being assembled from spare parts from other spacecraft; it now seemed to be held together by baling wire and glue.

Though the single strings were all worrying, they would not immediately put the spacecraft out of action. But the spacecraft was in danger of overheating badly—something that had begun to worry the flight controllers on the cruise to Venus, and the problem had been getting worse ever since. The cone-shaped medium-gain antenna, which fortunately was impervious to the heat, reached 100° Centigrade—the temperature of boiling water on Earth. So far, the flight controllers had prevented the overheating of the delicate equipment inside the spacecraft by turning the big antenna toward the sun and using it as a parasol—a maneuver they called hiding. (The medium-gain antenna stuck out too far to the side to benefit from the parasol.) The spacecraft computer could not be allowed to get hotter than 55° Centigrade, and it did not. And there were other heat-sensitive systems all over the craft—the most sensitive were the batteries, which could not be allowed to get above 28°. No temperature inside the craft ever got above 36° and the batteries were kept cooler.

The overheating worry was caused by the fact that the little square mirrors which covered much of the insulation to reflect the sun's heat away, and which had never been used on a mission before, were degrading. Space engineers, who regularly deal with problems an astronomical unit or more away in spacecraft they can't see or touch, have developed a working method of thinking of all the possible alternatives to a problem and then simulating them on Earth to see which comes closest to duplicating the symptoms; they seem to vie with each other in producing alternatives. So far, they had thought up four possible explanations for the deterioration of the mirrors: either the black material at their backs had degraded so that they no longer reflected properly, or the particles of Astroquarz that were flaking off the insulation were slowly snowing onto the mirrors and covering them up, or some sort of contaminant from the space shuttle had settled on the mirrors while it was still in the cargo bay, or the silicon-rubber glue that attached the backs of the mirrors onto the spacecraft had contaminated their fronts—silicon-rubber is a clear fluid, but in space, over time, the sunlight would darken it. Whatever the case, there was no way to fix the mirrors. Instead, measures had to be taken twice each orbit to shade the spacecraft by using the big antenna as a parasol, and whenever that happened (the engineers called it the double-hide maneuver) the spacecraft could neither map nor transmit data to Earth. The problem would be worse in different parts of Venus's orbits around the sun, and accordingly the second, fourth, and sixth eight-month mapping cycles would be hotter than the first, third, and fifth, and would require more shading; and the problem would get worse as the mirrors continued to deteriorate. As the periods needed for cooling got longer, they would cut increasingly into the playback time—and hence into the mapping time; currently, the length of each noodle was being cut by more than fifty percent.

And there were problems on the ground that were just as serious a threat as the problems in space. Funding was a continual annual rat race. On projects like Magellan, the mission managers never knew more than a year ahead how much money they would have, and each year they devoted much of their energy to fighting the battle for the next year's budget. For fiscal year 1991 (which began in October 1990), Spear had been told he had $43 million, which was on the slim side of what he needed. (As a rough rule of thumb, it cost $15 million annually to operate the spacecraft, a second $15 million to gather and transmit the radar data, and a third $15 million to process the data and pay the scientists to analyze it—a total of $45 million.) Much of the $43 million that Spear was getting had been reallocated from other planetary projects, and it wasn't enough. Earlier, Spear had been told by NASA headquarters that another $5 million would very likely be found for Magellan, and the project was counting on it, but he had recently been told by NASA headquarters that that money could not be found—he only had the $43 million. Economies would have to be made, and Spear did not know where—in the data processing, perhaps, or in the scientists. Clearly, the dangers to a mission lie as much in Washington as in space. Spear, who spent about half his time on budgetary matters, had once told me that he was delighted when Magellan was launched aboard the shuttle six months before its best window of opportunity, even though it meant taking a far longer route to the planet, "because it got us away from the threat of budget cuts." (The best time for a direct trajectory to Venus was in November 1990, but Magellan gave up its place to Galileo, which went to Venus as the first leg on its far longer and more risky flight to Jupiter; hence, Magellan took an earlier flight, though it meant arriving at Venus later.) The sacrifice of Magellan's berth and the earlier departure had gained Spear nothing; he had not realized that the budget cuts had rockets of their own and would pursue Magellan all the way to Venus.

Already, Spear was receiving anxious messages from scientists and engineers who feared the ax. At the same time as the funding was diminishing, he had to ask for additional funding for the mapping cycles following the first, which would end in the middle of May. NASA budgets only a spacecraft's primary mission—in Magellan's case, the first cycle. "When we plan a primary mission, we put in extra capacity so that we'll be sure of success," he told me. "Then, if we are successful, the extra margin is still there, for an extended mission. Extended missions are like gravy—the most wonderful example is Voyager, whose primary mission was Jupiter and Saturn, but which made it to Uranus and Neptune as well, despite threats to terminate the mission after Saturn.

"But asking for an extended mission is like asking for a brand-new mission. We have to go to the end of the funding queue. And we are unpopular, because for years Magellan has exceeded its budgets and taken money away from other projects. We have to think of things to justify the future cycles—like filling in the gaps in the first-cycle imagery, or looking at the surface from a different angle, which will provide different information about the surface, or provide stereo, or looking for changes such as volcanism from one cycle to the next, or the aerobraking for the gravity measurements.

"Congress is easy to tap for money if you have a disaster, such as the Challenger or the space telescope, but if you have a working spacecraft—there's no

money! But I suspect we'll get that money, because
NASA never has turned off a mission prematurely.
They won't turn us off, but they may turn down the
spigot. The rate of processing can be cut. The rate
of acquisition can be cut. The size of the science team
might be reduced." He felt rotten about it, he said,
because the morale of the Magellan scientists and
engineers, which had been superb, was beginning to
degrade as clearly as the mirrors on the spacecraft.

When I ran into Okerson a little later, he suggested
that the project ought to suspend the nomenclature
committee and sell naming opportunities for craters
and other features on Venus, as is done in major
fund-raising campaigns for new buildings; and
though he had the deadpan manner of someone who
thought he was being very funny, I had the feeling
that he also thought he was on to something.

IN the middle of March, I went to the twenty-second
annual Lunar and Planetary Science Conference at
the Johnson Space Center in Houston. On the first
day, I ran into Tony Spear, who had a badgered but
triumphant look. The Magellan scientists were at the
conference in force, and they were in a bad mood
because of the threat of cutting funds for science or
processing. The scientists had held a stormy team
meeting in Houston at which they protested the cuts,
and they also attacked the project for delays in
processing the data, which for a variety of reasons
other than funding was far behind. Only one of the
hundred-odd compact disks of data containing images

of the first cycle had been released, and the scientists were muttering that the project was not living up to its commitments.

The threat of cuts, and not knowing where the cuts would fall, had sent bad vibes through the engineering side of the project, too. Spear is easily made miserable by bad vibes. Just before the Houston meeting, Spear collected all his key managers and took them on a three-day retreat at a hotel at Newport Beach. He told them that he was tired of taking flak because of the reduced funds, and he asked them to figure out where the cuts should be made. "For the next three days, we focused on nothing but the budget," Spear told me. "They realized there had to be cuts, and they involved themselves thoroughly in the process. I learned I could trust them to do the right thing. And they loved it! We all had a wonderful time. I learned a big lesson in management. We're going on another retreat when we have to discuss the 1993 budget."

Spear was able to announce to the angry scientists in Houston that Magellan's deficit problem had been largely overcome—enough money had been found, in part because Martin Marietta had been persuaded to accept a delay in receiving some awards money due from NASA, in part through cutting funds for speeding up data processing, and in part by cutting funds on the engineering side, particularly funds for trouble-shooting and problem solving. Consequently, the funds for science would not be cut at all. He had also been able to report that the bugs in the processing system had been largely ironed out.

Because the squeeze was over, at least for the

moment, Spear was in a more cheerful mood later that day than when I had first seen him. "There's going to be a good view of Venus tonight," he said. (Spear seemed to spend more time looking at Venus—the real Venus, as opposed to its radar images—than anyone else connected with Magellan.) "Venus, Jupiter, and the moon will be close together just after sunset. Whenever I look at Venus, I can almost see the data bits streaming from there to Earth. Venus is now a little farther from us than the sun, and I figured out last week, when I was looking at it from my back lawn in Pasadena, that there is one bit per mile." That night, as I went with a couple of Magellan engineers to a fish house on Galveston Bay, we stopped to admire the bright crescent moon, with a faint dot and a brighter, more luminous dot just below it—like an illuminated mobile hanging in the indigo sky. We were unable to make out the stream of bits, however.

SAUNDERS had urged me to go to the Houston meeting. "Here at J.P.L., we're all cooped up, with the data pouring in and with no one to talk to except ourselves—and most of the team isn't here that much," he had told me. "I really enjoy going to meetings of professional scientific groups, where we present the data and our preliminary ideas—and then listen to what the rest of the people in the scientific community have to say." It was at such conferences that some members of the team found out for the first time what their own teammates were doing. "At Pasadena, we're all so immersed in our own work that we don't pay attention to what others are doing—and we're not always at Pasadena at the same time," John Guest told me. "At professional conferences, I find out what the Magellan tectonics people are up to, and they learn what we volcanologists are doing."

The Houston meeting lasted five days, and the Magellan scientists presented their papers in the first day and a half. About a thousand planetary scientists were milling around in a large conference center, dashing in and out of simultaneously running sessions—the conference center had two gymnasium-sized auditoriums and several smaller lecture rooms, and it was about as easy to take it all in as a five-ring circus. I attended most of the Magellan scientists' talks, and sometimes the feedback came in the question period after each talk. At the end of Arvidson's talk about the horseshoes, Eugene Shoemaker, who, although he was not a member of the team, had kept up with the Magellan data through such Flagstaff colleagues as Schaber and Soderblom, got to his feet and asked if there might not be what he called a source problem—that is, could all the horseshoe ejecta, which sometimes spread across a thousand miles, actually be stuffed back into the crater it was supposed to have come from? Arvidson replied that he thought it could, if it was no more than a few centimeters thick. Shoemaker himself received a bit of feedback from Phillips, who at the end of his own talk about cratering told Shoemaker that he should now go back and refine his estimates about asteroid and cometary impacts on Venus in light of the new

evidence from Magellan—his revisions were eagerly
awaited. And at the end of Basilevsky's talk about
the likelihood that Venera 8, which had detected an
unusually complex granitic-type of rock, had landed
on a pancake, a member of the audience rose and
asked him what other interpretation he had for the
Venera 8 data, assuming the spacecraft had not
landed on a pancake. Basilevsky replied that the
imagery from several of the Venera landers had
shown layered, tuff-like rocks (such as he had told
me about earlier), and that tuffs might also account
for the complex chemistry detected by Venera 8.
When I saw Basilevsky the next day over coffee in a
big common room at the back of the second floor,
he said, "During my presentation I was full of eu-
phoria when I showed the slide of the pancake inside
Venera 8's landing ellipse, but then this guy asked
that question, I had not thought of that before! So I
think the paper I am writing needs a couple of extra
paragraphs about the possibility of tuff." Head, at the
end of his own talk in which he had waxed eloquent
about the extraordinary detail of the Magellan imagery, had said he hoped the Soviets would now send
more landers to Venus, and I asked Basilevsky about
the possibility of this. Basilevsky said he did not think
the Soviets would be sending more Veneras to Venus
any time soon, but if they did, and if he was on the
targeting team again, he would target a tessera region,
a region of high reflectivity on a mountaintop, and
another pancake.

Possibly because of the frantic nature of four si-
multaneous programs, and the constant indecision
about which program to attend, a large number of

scientists were continually repairing to the big common room, where there were capacious coffee urns and dozens of tables surrounded by chairs. It is around the coffee tables that the true business of most conferences takes place anyway, and where a lot of the feedback occurs. Arvidson ran into a geologist from the University of Hawaii, Bruce Campbell, who set him thinking. Arvidson and Phillips are interested in the ages of the lava flows covering the plains. Campbell had developed a method of dating flows on Mauna Loa by their radar reflectivity. Young flows were jagged and highly reflective, but as the flows got older, the rough spots were filled with windblown dirt and became smoother, so that they appeared darker on radar. By comparing the radar with the known ages of some of the flows, he was able to date other flows. Arvidson got into a long conversation with Campbell about the chances of using the method on Venus. Solomon found the feedback of terrestrial geologists like Campbell particularly useful. "On Venus, we see all these ridge belts and groove belts, and we talk about compression or extension of the lithosphere, but we don't ever visit any," he told me. "There's a large community of Earth geologists who tramp around terrestrial features like these and have ideas about them, and they can be very helpful to us."

At another table, I found Peter Schultz, the cratering expert, with another Soviet geologist, Boris Ivanov of the Schmidt Institute of the Physics of the Earth —who was collaborating with Basilevsky, Schaber, and Kirk on Cleopatra—deep in conversation over how an asteroid plunging through the atmosphere of Venus changes its shape. Ivanov had been convinced by some work he had done a few years earlier that it flattened out a little like a button, but some work Schultz had done indicated that it would become somewhat bullet-shaped—the idea was that under the stress, it would take on a more aerodynamic form, following morphologically the course of least resistance. (When the stresses become too great to absorb them in this fashion, Schultz told me, they can cause the asteroid to break apart, in which case the largest pieces will continue downrange farther than the smaller ones, each making its own crater.) Ivanov now told Schultz that after further work he had discarded the button model and now agreed with Schultz about the bullet shape. After Ivanov had disappeared to get some more coffee, Schultz told me: "A lot of work I've done on the effects of atmospheres on cratering I've done on Mars and Earth. I was very firm on the idea of atmospheric effects on cratering long before anyone else was—how it modified the ejecta into flows and petal-like or butterfly-shaped features. Nobody paid any attention! But now, with Magellan, all these effects are much more obvious. And Magellan makes a lot of things I've been saying for years more acceptable. A little while ago, Mike Carr, who had been the leader of the imaging team for the Viking orbiter, came up to me and said, 'Gee, maybe you've been right all along.' "

When Schultz went off to get some coffee, Ivanov returned, and I asked him his impressions of the Magellan data. He replied that he was delighted with the data he had seen here and the one CD he had received in Moscow; the trouble was his institute didn't have a computer capable of reading it—the only computer that could was one in Basilevsky's laboratory at the Vernadsky. Fortunately, he said, they are good friends. NASA is trying to help the Soviet scientists get more computers—and to organize their own archives of planetary material on CDs in the same format, so that scientists everywhere can use both countries' material more readily.

Downstairs, at the back of the largest lecture room where the Magellan scientists were speaking, there were several rows of freestanding bulletin boards where a number of scientists who were not giving talks had tacked up what they called poster presentations—small exhibits about the research they were doing. A great many of them were done by the Magellan graduate students, including Kari Magee-Roberts with her map of the finger flows at Mylitta Fluctus, and Brennan Klose, who had put together a display about the pancakes, in which he seemed by now to have a proprietary interest. The pancakes, which are eye-catching, were drawing more attention than the flows, and much of the time Klose was surrounded by a small crowd. Klose looked as if he was suffering from an overabundance of feedback. Just a little earlier, Peter Ford had delivered his talk about the altimeter data giving the same altitude reading at the center of the pancakes as in the surrounding plain—suggesting that they were not pancakes at all but English muffins. (The error in processing the altimeter data had not been discovered yet.) "It's hard to believe, because they look higher at the center," Klose told me after one group of scientists had left and before the next had formed.

....

"I've just had a useful talk here with John Fink, the geologist who knows the most about the pancakes on Earth. He says that the centers are not honeycomb material, which Ford suggests is the explanation for the fact that the altimeter sees through them, but that they have a lot of gas bubbles in them, and that might account for Ford's readings on Venus." In the flip-flop over the pancakes and the altimeter data, the arguments over bubbles versus honeycombs melted away, and several prospective papers with them. There would be other errors, too—volcanoes that appeared where no volcano had been, and then disappeared as rapidly as the data were reprocessed; appearing and disappearing landslides and sand dunes. Ingenious theories that explain the initial observations arise with great ease (such as the bubbles or the honeycomb inside the pancakes) and then disappear—or (as in the case of the center of Cleopatra) are revived and revised. Then wry jokes are made about how scientists can concoct a theory to explain any phenomenon—whether the phenomenon is real or not.

Upstairs, in the coffee room, I ran into an Australian geochemist, S. Ross Taylor, whom I had known for many years; an expert in the crust of the earth, he had done the first chemical analyses of the rocks the Apollo astronauts brought back from the moon in 1969, spoke at the first of the series of lunar and planetary science conferences in January 1970, and presented papers at all twenty-one conferences since. Taylor, who is a large badger of a man in his middle sixties, told me he was currently writing a book about the crusts of all the solid inner planets, including our

....

moon. I asked him what he thought were the chances of granitic material in the crust of Venus. "I've been thinking about that question for my book—that's one reason why I'm here," Taylor told me. "The basic message of the book will be that every planet and moon is different from every other. Magellan is giving us an unbelievably beautiful view of Venus, and I think I can build up a consistent picture of it—what happens to a planet when you push up enormous amounts of basalt. I'm fascinated by the coronae and the circular regions! But it doesn't look as though the planet has much granite, because it doesn't appear to have had plate tectonics. Granites require many episodes of subduction and remelting, and you're not going to get that on Venus. And there's no water. Even without water, you'll always get a little granitic material crystallizing out of a melt—as it cools underground, the basalts and other, less complex minerals drop out first, and what's left at the end is a little pot of complex stuff that gets pushed to the surface. So I'm satisfied there might be enough granite or rhyolite for a few pancakes. But there will be nothing on a continental scale."

WHEN I saw Head a little later, he told me that the spreading question had not progressed much further than when I last saw him. By this time, the data from eastern Aphrodite, where Head had had the greatest hopes for rifting, with what he thought were the best examples of transverse faults and a central rise suggestive of crustal spreading, were beginning to be processed. As he had been by Ovda, Head was amazed by the complexity of the surface in eastern Aphrodite. "You look for unifying ideas— but you are overwhelmed by the complexity," he said. "There is regularity and complexity. There are linear fractures and ridges running parallel to and cutting across the trench—as you might expect from spreading. But there are many more! It gets really confused. There are distorted and contorted zones. You see lots of volcanism, and lots of deformation. There are impact craters there. But you can't say whether they are in excess of what you'd expect for a young spreading center. When we proposed the crustal-spreading idea, we were fully aware that there would be deformation, because we suspected the crust was somewhat plastic. But it's much more distorted than we thought. When I described these ideas to David Strangway, a geophysicist at the University of British Columbia who is an expert on seafloor spreading, he said if you were walking on our ocean bottom you would not know that plate tectonics was going on. We're so used to the maps of the ocean basins, which were made by sonar and show all the rifts and the faults, that we think they would be simple to see. But the ocean bottom isn't like that, visually. I've looked at pictures of the East Pacific rise crest taken with a towed submarine vehicle—there's a major rift there —and it's hard to make it out. Those vehicles are analogous to the Magellan spacecraft. At this point, I'm still not convinced about any of these ideas, but I can't rule any of them out." What one might or might not see on a stroll through the muck and ooze and sand dunes of the abyssal plains if Earth's oceans were drained, and what one might or might not expect to see on the plains of Venus if they were like our ocean bottoms, I was finding, was a subject of some disagreement.

Most of the other scientists, however, did not feel that the complexities of Venus might conceal Head's spreading. A little later, Solomon told me again, as he had after Ovda, "Head's ideas are dead. The specific transform faults, the crater densities, and the rises that he said would be in eastern Aphrodite are not there." A certain degree of asperity had entered the dispute, and not all the scientists were inclined to be magnanimous in what they took to be their victory. Bindschadler felt that Head had ignored some important papers that had shown that spreading was not occurring on Venus—one by Phillips and Kaula of U.C.L.A. in 1981 and another in 1989 by Solomon and Robert Grimm (the ponytailed post-doctorate who earlier had been running the computer for the crater group). There seemed to be a certain element of self-righteousness on the part of geophysicists toward a geologist who had trespassed too far onto their turf of planetary interiors and was getting his just deserts. But the feeling of vindication was by no means limited to the geophysicists. Jerry Schaber, who himself had written a paper in 1981 suggesting that there were rifts on Venus of limited extension, felt that Head had pushed the idea of rifts as spreading centers too far and held on to it too long. Phillips even accused Head of writing too many papers about spreading just prior to Magellan's arrival at Venus. In part as a result of the bad feeling (and in part also because the geophysicists felt that tectonics, reflecting

as it does internal processes, was more in their preserve than in that of the geologists, the same sort of
argument they had made about the craters), Head
was pushed out of his work with tectonics—a difficult
situation, as he was chairman of the geology- and
tectonic-processes working group. The group split in
two, with Solomon retaining the tectonics area and
Head becoming chairman of a new volcanology working group, which included John Guest's old group.

I thought Head's colleagues were being a little hard
on him. Head is not the first scientist to persist in his
ideas longer than his opponents felt he should or to
fail to give as much weight to papers that disagreed
with him as his opponents believed he ought. Nor is
he the first to promote his own ideas in scientific
journals, which decide for themselves what to publish,
usually after peer review. One hidden element in all
this was the question of grants and funding—the
competition for funding underlay much of the scientists' maneuverings (and indeed the maneuverings
of the Magellan project as a whole in relation to
NASA) as surely as the mantle with all its upwellings
and downwellings underlay the lithosphere. In this
arena, having a theory which later proves correct can
be translated into grants—and the converse can also
be true.

But I wondered whether in this case there might
not be another element—the lack of collegiality on
the team. Not only were the scientists a less cohesive
group than on most missions, because they were so
seldom together, but they were constantly in a state
of irritation about data processing as well as funding.
And in Head's particular case, one scientist told me,

there was a certain envy of his phalanx of (to all appearances, at least) well-subsidized graduate students—who, some feared, would snaffle up the data and make all the discoveries. (Reportedly, some of the Brown students were so turned off by what they felt was hostility toward them that they wondered whether they wanted to become planetary scientists at all.) One possible ramification was the fact that during the early part of the mission Crumpler, Head's post-doctoral research associate and co-author of some of Head's pre-Magellan papers on spreading, was not allowed to work with the team at J.P.L. Officially, the science team felt that Crumpler would get access to the data through Head; however, Crumpler felt that this restriction was not applied universally, and it was later reversed when other non-team members began to visit J.P.L. and work with the data. It's no news, of course, that politics enters science, as it does all other fields of human endeavor.

When the scientists were in a better frame of mind, they recognized that Head and Crumpler's ideas had considerable value. The spreading theory had remained a possibility until Magellan imaged Aphrodite—even Phillips had complimented Head on producing a theory that made certain predictions whose confirmation could be looked for in the Magellan imagery. Saunders had told me it was useful to have Head's spreading theory—it provided a framework for the scientists to organize their ideas upon. "A lot of scientists do their best work when they are trying to prove So-and-so is wrong," Bindschadler said, suggesting that Solomon might have been so inspired in his 1989 paper. Basilevsky said to me, "Sure, it was useful for Jim Head to have his ideas. If scientists have only one idea to talk about, there is no discussion, and that stops science." (Despite their professional differences, Basilevsky and Head have remained close friends; Head invited Basilevsky to spend the academic year 1992–93 at Brown, where he currently is. Solomon and Head have also remained personally on good terms.)

Head himself showed no signs of dismay when I talked to him a few weeks later; he was not thinking in terms of victory and defeat, or right and wrong. "I don't gravitate toward extreme positions. I have ideas, propose conditions that have to be met," he said. "The way I work is to say, 'This is a viable hypothesis. Let's test it.'" Perhaps more than his colleagues, Head has the type of mind that can become enthusiastic about one possibility while keeping other possibilities well in view; enthusiasm need not rule out objectivity. Later in the spring, Head himself began moving away from the spreading theory—it just took longer for the complexities of eastern Aphrodite to cease beguiling him. What pushed him hardest, he told me afterward, were the cratering rates that showed the surface ages were quite similar—something, of course, that had unsettled him as soon as he had heard it. However, despite his objective stance, Head would not have been human if he hadn't been disappointed at the way the data appeared to have overtaken the spreading theory, which had occupied his attention for several years; to his credit, he threw himself into his new work with volcanoes.

The ideas about plate tectonics were hard to get

rid of, though; and some scientists were arguing that whether or not there was spreading or plate tectonics on Venus today, the planet might have had them at some point in the past. Perhaps there had been a time, before the greenhouse effect had set in, when Venus's atmosphere had been thinner and much of its carbon dioxide was still locked up in its interior, when the surface temperature had been cooler and the crust more rigid. Then water could have existed on the surface in possibly oceanic abundances, sufficient for an asthenosphere that would lubricate plate tectonics. What lent credence to these ideas was a finding by Pioneer 13, an American spacecraft which in 1978 dropped into the atmosphere a probe carrying a mass spectrometer. By accident, a droplet of sulphuric acid (composed of hydrogen, oxygen, and sulphur) from a Venusian cloud got into the instrument, which analyzed the gas and found an enrichment in deuterium, a heavy isotope of hydrogen present in water in tiny amounts, a hundred times greater than would be expected if there had never been any water on Venus. From this, scientists deduced there might once have been enough water to make an eight-meter layer all around the planet. However, this theory was controversial because of the accidental nature of the measurement. Even if the measurement could be verified, many scientists refused to predicate a global sea on a few molecules of a hydrogen isotope.

Though scientists often cannot rule out a theory, they may be able to impose more and more constraints upon it so that it looks increasingly unlikely, and this seemed to be the case with plate tectonics in an earlier

Venusian epoch. The time could not have been more recent than 800 million years, now thought to be the age of the oldest part of Venus's present surface. And that upper limit on the latest possible date was being pushed even further back. Solomon told me of a theory of a Japanese geophysicist, Takafumi Matsui of the University of Tokyo, which concerned the decay inside Earth (and presumably inside Venus, too) of a radioactive isotope of potassium, potassium-40, into an isotope of argon, argon-40. The most efficient way of getting the argon-40 from the interior of the planet into the atmosphere, where it can be measured, is by plate tectonics, Solomon told me. It so happens that there is only about a quarter the percentage of argon-40 in Venus's atmosphere as in Earth's—which Matsui took as suggesting that if there had been plate tectonics on Venus in the past, it would have to have been in the first two billion years of the planet's history.

Ross Taylor was not content to let the question lie even at that remote epoch; he felt he could push the date back virtually to the birth of the planet. The deuterium in the sulphuric-acid droplet did not impress him. He told me, "As I'm almost certain Magellan has revealed no continents on Venus, there can never have been subduction on a major scale or plate tectonics, because continents require granite, and once you have light granite continents, you can never get rid of them." Though there were some scientists, such as Head and Klose, who were not yet willing to rule out large amounts of granite on Venus—even continental quantities—a great many of the scientists agreed with Taylor.

The scientists, Head prominent among them, were becoming less concerned with pre-Magellan theories and more interested in what the Magellan data themselves were saying about Venus and Earth. Saunders had once said to me that in the study of any planet there is first a tendency to apply ideas from Earth to the planet, and later, when enough is known about the place so that the scientists have grasped its unearthly individuality, they begin to apply their new discoveries in the reverse direction—as was the case with the splotches. The transition seemed to be occurring now at a greater rate. "There is a lot of speculation about when seafloor spreading started on the earth," Phillips told me. "Maybe the earth's lithosphere was a lot thinner in its early days, up to the end of what we call the Archean period, two and a half billion years ago. Perhaps the earth's lithosphere was hotter then. One concept is that if the earth's lithosphere was hotter or thinner in its early history, you could not have had rigid plates or plate tectonics. You might have had a more crumply, messy, scummy situation—such as we see on Venus." Conversely, Venus today may resemble the future earth, several billion years from now. A geophysicist at Stanford University, Norman Sleep—one of the expanding circle outside the science team to make use of the Magellan data—told a writer for *Scientific American* that in the future Earth's surface might be dominated by upwellings and downwellings, the way Venus's is today. He speculated that the more plastic crust on Venus had allowed its interior to cool more rapidly than Earth's; and that when Earth has cooled to the same extent, plate tectonics will stop and the plumes will take over as the major force influencing the surface, with Ovda and Beta Regiones, Ishtars and Maxwells, popping up in the Atlantic and Pacific oceans, the Arctic and the Indian.

Missions like Magellan, which answer many questions, often raise many others. "One of the problems is that nobody agrees about what makes the present system work the way it does on Earth," Phillips said to me. "For example, we don't even know if water is essential to plate tectonics, even though it certainly contributes to the process on Earth. You could argue that if there were no asthenosphere, the coupling of the plumes directly to the lithosphere might allow the plates to keep on chugging. But the plume theory itself is even less well worked out for Earth than plate tectonics is—some scientists are saying, for example, that plumes rising from the core-mantle boundary wouldn't get beyond the discontinuity farther up the mantle. And so one of the difficulties of working with Venus is that we don't yet have a good model for Earth. In fact, that's the beauty of the whole thing. We now have two planets to deal with—and that gives us a bigger observation span."

PART III

A Pretty Bizarre Thing

IN May, many of the problems that had plagued the engineers began to be resolved. To begin with, as there had been no more alerts in memory B since the barrage of fifty-seven in November, the engineers now decided that the alerts had definitely been caused by a hardware problem—most likely, the theory had been correct that a silicon chip had been damaged by an electrical charge at the time of the separation of the rocket after the VOI burn, and that it had indeed by now healed itself. Consequently, they reprogrammed memory B—which had been filled with trap instructions—and gingerly brought it back on line.

Though Rick Kasuda, the leader of the attitude-control software team at Denver, and Bob Reilly, a member of the team, had been taken off the LOS problem at the end of October, two months after the second LOS, they couldn't help looking at the delayed-engineering data after the third LOS in November 1990—the one I had been present for. There had been a fourth LOS on March 5, 1991, two weeks before the Houston conference, and they took a look at the data from that, too. There was a fifth LOS on May 10. In the last two cases, the IODAs were swapped, and the spacecraft was back in business quickly. After the March LOS, the spacecraft suffered two hours of total loss of communications, though it took another four hours to tweak the attitude to its proper position. The May LOS took longer, with six hours of total loss and twenty-four hours to tweak— the DSN did not have the right antenna to get in, as had happened at the start of the third LOS. Still, this was far better than the performance on the first two LOSes, and the flight engineers were pleased with their ability to control the problem—if not to solve it. They had not been able to prevent the erasure of the tapes on the later LOSes, and the memory read-outs hadn't helped much, either. Still, with the increase in the number of LOSes, there was an increase in the statistics. Reilly told me later, "I didn't discover anything new after the third or the fourth LOS, but after the fifth one, I knew it must be a problem in the software for certain tasks." Of the five LOSes, three of them—the first, fourth, and fifth—had occurred during a star cal, and one of them, the third (for which I had been present), had occurred during mapping. Since three of the five came during star cals, Reilly turned his attention to the software for that task.

Reilly's attention had already been focused on the star-cal software for four days before the fifth LOS, because it had been implicated in an event in the systems-verification laboratory at Denver, where the engineers continuously ran computers identical to Magellan's, loaded with the same commands that they sent to the spacecraft. On May 6, right in the middle of a simulated star cal, one of the computers had an

RPE. In one key respect, the computers in the SVL were set up differently from the ones aboard the spacecraft: when the heartbeat failure occurred, the memories were swapped—not the IODAs. For this reason, the information leading up to the RPE was preserved further back than it had been after any of the RPEs in the spacecraft. The SVL engineers took two weeks to analyze the situation—maddeningly, the fifth LOS occurred before Reilly, who was looking over their shoulders, thought of the probable solution, though the fifth LOS convinced Reilly that with the star-cal software he was on the right track. He got the answer several days later, sitting at his desk in Denver. Immediately, he went to look for Kasuda and bounced the idea off him. Kasuda liked the idea. He searched the SVL memory's delayed-engineering data and found what Reilly had suggested he would find.

The answer involved taking the step into the next level of detail, which Okerson had written about in his memo to headquarters on August 31, 1990—when he (reporting Kasuda's thinking) had talked of evidence that the processor had suffered cycle slipping as a result of a failure to complete "each 30-Hertz cycle prior to the arrival of the next interrupt," and the "watchdog timer" had failed to go off. I asked Okerson about this later—before I met Reilly and Kasuda.

Like the human brain, Okerson told me, the processor does a great many things at once, but unlike the brain it does them by what computer technicians call multiplexing a variety of tasks—that is, it breaks them up into tiny segments and interleaves them, going from one task segment to another in a cycle very rapidly. While it is carrying out a star cal, it may also be computing information it will need for an upcoming maneuver, and monitoring every aspect of the spacecraft—the current being generated by the solar panels, the current coming out of the batteries, the temperatures of a variety of systems, or the pressures in various fuel tanks. Activities like attitude changes for the star cal, which are time-critical, are called foreground tasks, and activities like monitoring or computing, which need not be done immediately, are called background tasks; there are six different ones. The processor cycles through all the spacecraft's foreground tasks, and some of its background tasks, thirty times a second—that is, every thirtieth of a second, it pushes each of the foreground tasks a little further along, starting with the more important ones like conducting the star cal. And if there is time left over, it goes on to as many of the background tasks as it can get through before the cycle ends. The end of each cycle is marked by a signal, called the 30-Hertz signal; whenever it arrives, it interrupts whatever background task is going on and the computer goes back to the beginning of the cycle, starting again with the most important foreground task. The foreground tasks are done in discrete, self-contained units which all together fit entirely within the cycle; but a single background task may take several cycles to complete, and the segments are not self-contained but rather are part of a continuous flow which can be, and is, interrupted arbitrarily by the 30-Hertz timer at any point and picked up at that point in the next cycle.

....

The mechanism for regulating all this is delicate and complex—and if it breaks down, there can be chaos. At the start of each cycle, the processor notes the time, and after it has completed the foreground task, it checks the time again. Typically, the foreground tasks together take up twenty-thousandths of a second, allowing thirteen-thousandths of a second for the background tasks before the next 30-Hertz interrupt. If because of some problem the processor hasn't completed the foreground task before the end of the cycle, it overrides the interrupt and notes its tardiness in a spot in the memory called the cycle-slip counter. And one clue that Kasuda had noticed when he read the memory back after the first LOS was that the cycle counter had slipped by an enormous amount—that is, there had been a huge number of uncompleted foreground tasks. What had happened was that when the processor noted the time at the start of each cycle, the time was stuck at the same instant, cycle after cycle; but when it noted the time at the end of the foreground task, that time was correct. In very short order, the time in which the background tasks were supposedly being done stretched to an impossibly long period—whenever a cycle started, it was time to end it—so that none of the tasks, foreground or background, was being completed within the time allotted to it, and the cycle-slip counter raced upward like the meter of an orbiting taxicab.

Because of all the cycle slipping, the watchdog timer should have given the alarm, but it did not. The watchdog timer is set to give an alarm every few seconds—but if the spacecraft is working properly,

....

the software program will reset it so that it won't go off for another few seconds. Ideally, if the spacecraft is working properly, the watchdog timer will never go off, for it is continually being reset. If there is a problem in the processor or in the memory, the timer won't be reset, and there will be an alarm signal. But there wasn't—despite the cycle slipping, the alarm continued to be reset. "It was like the mystery which Sherlock Holmes solved when he noticed that the dog had not barked," Okerson told me. When Kasuda asked himself why this should be, he decided that the processor was caught in a tight loop—a runaway program execution, or RPE—in a part of the software that included the watchdog-timer reset program and the 30-Hertz timer. "The processor was just going around and around, setting off the 30-Hertz timer and resetting the watchdog timer, and doing very little else," Kasuda told me later. "I figured that out in August 1990, right after the second LOS, but it made no sense at the time." He and the other engineers didn't know how the computer got caught in that tight loop, or why. And the same two symptoms, it subsequently turned out, were common to four of the five LOSes—all except the second.

Unanswered questions on one level of a computer often require another step down to a still lower level of detail, and Reilly, Kasuda, and their colleagues at Denver took this step in May, after the RPE occurred in the systems-verification laboratory. The question came down to the manner in which the processor knows whether a background task needs to be picked up on the next 30-Hertz cycle and how it leaves the task each time it puts it aside during the last cycle, so that on the next cycle it can be picked up again at the right point. Each task starts and ends with what engineers call a flag (actually, a binary code number), which says one of two things: either the task is "Active," or it is "Scheduled." A background task is initiated when the software for a foreground task decides it needs to have it run—for instance, an attitude-change foreground task might need a background task to calculate where a particular guide star is in relation to the spacecraft, and hence the foreground-task software in the processor will ask that a "Scheduled" flag be put on that background task, so that that task will be picked up in the next cycle when there is time available. The flag is put in place by a part of the processor called the background-task scheduler. The computer will start a task flagged "Scheduled" from scratch; when it is started, the software for the background task raises a second flag that says it is "Active." The two flags are up together.

If a background task is interrupted *in medias res* at the end of the 30-Hertz cycle, a record of the precise point at which it stopped is kept in a part of the memory called the stack. Both flags remain up, signaling the processor that the task is incomplete and should be picked up again during the next 30-Hertz cycle that has the time for it. Triggered by the two flags, the processor will go to the stack to retrieve the five or six words that will tell it precisely at what point to pick up the task—computer engineers call retrieving this information popping the stack. Okerson said to me, "Think of a background task as similar to trying to follow a very long recipe in a cookbook while you are doing ten other things, so that you are constantly closing the book and putting it aside in the middle of your cooking, and each time you leave a note to yourself saying, for example, 'Start on p. 516.' Then you do the ten other things, and you come back and read a few more pages of the recipe, and leave another note that says, 'Start on p. 521,' and so on, and on. When you are done with it, you close the book and put it away." When the background task is finally finished, two-millionths of a second later both flags should be lowered. In Magellan, the task scheduler in the processor lowered the "Scheduled" flag but the background task's own software lowered the "Active" flag.

This is what was supposed to happen—and what did happen on almost all occasions. Reilly told me, "In May, after the RPE in the SVL, I asked myself, 'What if in the two-millionths of a second after the task was complete, something intervened? Say you got a 30-Hertz interrupt. There would be a problem.' I bounced it off some other people, including Kasuda, and they agreed that could happen."

With the evidence pointing to the star cals, Reilly suspected the particular background task in which the computer has to compute the precise position of a guide star, something that has to be done twice, as each star cal requires sightings on two guide stars. "I took Bob Reilly's speculation and proved with the memory in the SVL after the RPE that he was right, and that what he suspected had very likely happened aboard the spacecraft," Kasuda told me. What Kasuda discovered when he examined the memory was that after the first of a pair of guide-star sightings had been completed, the task had been left with the wrong flagging: the "Scheduled" flag was down but the "Active" flag was still up. Evidently, the 30-Hertz interrupt had in fact arrived in the middle of the two-millionths of a second window of vulnerability. (The smallness of the window explains why the LOSes happened at rare and random intervals.) The background-task scheduler responsible for lowering the "Scheduled" flag had the power to override the interrupt, but the task itself, responsible for lowering the "Active" flag, did not have that ability. "No one ever realized it was possible for an interrupt to arrive at that point, so that a completed background task would be left with the 'Active' flag up and the 'Scheduled' flag down," Okerson said. "It made no sense in terms of the computer's own logic, for a background task cannot be 'Active' if it isn't also 'Scheduled.' "

This situation caused trouble minutes later when it was time for the processor to get started on the calibration with the second star. When the foreground task asked the task scheduler to raise the "Scheduled" flag on the same background task for computing the position of the next guide star, the scheduler saw that the task was already flying an "Active" flag. The processor mistakenly thought it was continuing an ongoing task. But when it went to the memory to pop from the stack the instructions about where to pick up the task, there were no instructions. As the processor was programmed to find instructions, it popped something else instead and went off and did that. The something else necessarily was irrelevant— what computer technicians call garbage. Okerson says it is what would happen if you reached for your cookbook but picked up, say, the World Almanac instead, and finding on the page number you had noted down the batting averages for the Baltimore Orioles, you tried to turn those into whatever you were cooking. Chaos would result.

In the case of the computer during the LOSes, it went off following what it thought were the right instructions, but went around to a lot of inappropriate addresses—it began running around a tight loop or runaway program execution very rapidly, hitting the watchdog-timer reset button and the 30-Hertz timer, causing the cycle-slip meter to race upward. (The same erroneous batting averages were popped from the stack each time, explaining the similarity of the symptoms from one LOS to another.) The computer would have continued in this cycle forever—except that the tight little loop also omitted initiating the heartbeat signal, which has to be done every two-thirds of a second. The resulting heartbeat failure, of course, is what triggered the hardware swap and the massive reset, the swift kick which knocked the computer back to its senses.

There was still the question of why the flags were vulnerable to the rare and random occurrence of the interrupts at just the wrong couple of microseconds. This involved yet another step down into an even more detailed level—and in this lowest level, the Denver technicians found none other than themselves. It was like Dorothy's discovery, when there was a sort of RPE in the land of Oz (no odder a place than the inside of the Magellan computer), of the Wizard himself behind a curtain, where he was pulling all the levers and creating all the difficulties; the

Wizard turned out to be someone very familiar who subsequently set everything right.

The Magellan computers have a relatively small amount of memory space, and therefore Bob Reilly, who had written the software four years earlier, had had to take some shortcuts to economize. In most spacecraft the background task does not lower its own "Active" flag. Instead, when a background task is finished, the task's software goes to the task scheduler in the processor—which has the power to override interrupts—and asks it to lower the "Active" flag, the way it did on Magellan for the "Scheduled" flag. The shortcut Reilly settled on, for the task to lower its own flag, was reviewed and approved by his teammates and superiors. "Though there is some argument about it, software philosophy today is, generally speaking, to put the knowledge about an activity within that activity's own software," Reilly told me later. "We wanted to put the knowledge of whether the background task was active or not active in the task's own software. As a shortcut, it was a good idea, and it conformed to theory. We tested it a lot. You'd expect a fault in the software to come out in tests—but the tests turned up no problem. We now know that events had to turn out just right for the fault to appear. It was a pretty bizarre thing." Reilly, who had written the programs, and Kasuda, who had joined the project later, were treated as heroes for their dogged (and unbudgeted) solution of the problem. They received commendations. Magellan provided a classic demonstration of the sort of problem analysis and crisis management that NASA at its best has always been known for.

The Galileo engineers, who had been looking with increasing anxiety over the shoulders of the Magellan engineers since the previous August, were able to relax. Galileo's computer had more memory space, and its programmers had not had to take any economizing shortcuts in the software. (Though Galileo is in the clear with respect to its computers, it has another problem—its high-gain antenna, furled at launch like an umbrella, would not open. If the Galileo flight controllers couldn't fix it, the amount of data Galileo could transmit could be drastically reduced.)

Everything—or almost everything—was set right on July 2, when the engineers uplinked new software to the spacecraft that eliminated the shortcut so that the background task would ask the task scheduler to lower the "Active" flag every time it was finished, as it already did with the "Scheduled" flag. As the same shortcut existed on all six of the background tasks, the software for all six was changed in the same way. Okerson and Slonski believe that this solution fixes the problems that led to four of the five LOSes—the first, the fourth, and the fifth, which happened during star cals; and the third, which happened at the end of a different background task during mapping. The one LOS which is not accounted for, and which Slonski and Okerson believe might recur, is the second, which happened during ROM; in ROM, the spacecraft does not do the tasks it does in RAM, and consequently the flagging problem is not an issue. The computer was also in ROM during the unexplained RPEs in the latter part of the first LOS. "Something—another computer problem—is out there which might yet eat our lunch," Slonski told me.

Some engineers such as Ledbetter suspect a hardware problem in memory B, which had come on line during the first LOS and which had had problems. Others—Slonski, Kasuda, and Reilly among them—suspect a hardware problem in processor B, the one that had not been used since launch but (like memory B) was brought on line when the entire AACS was swapped following the first heartbeat failure. It (again like memory B) had remained on duty throughout the first LOS, when a second heartbeat failure had caused the spacecraft inexplicably to exit RAM safing and enter ROM safing, and when a third heartbcat failure occurred shortly after that, also in ROM. When the first LOS was over, the engineers had left the spacecraft exactly the way it was, with processor B and memory B on line and in ROM, while they tried to figure things out. After the second LOS was over, processor B had been retired, the engineers hoped permanently—though there was always a chance that a weirdness might bring it back on line. "I suspect if that ever happens we'll be faced with another problem," Kasuda said. The engineers do not wish to hear from processor B ever again. Unlike memory B, there was no evidence that it had healed itself—if indeed there was anything wrong with it in the first place.

WHEN I went out to J.P.L. in November 1991 (a year and a quarter after VOI), I ran into Spear outside his office, hurrying off to a meeting; ever since the solution of the LOS problem the previous spring, he told me, Magellan had been running extremely smoothly. He said, "Magellan is so automatic now that it just rolls on—the only major problem is funding." There had been another funding flap over the summer when NASA had threatened to terminate Magellan at the end of its second cycle, on January 15, 1992, because of its need to find an extra billion dollars for its projected space station. Funding, however, was now secure through September 1992, the end of the third cycle—though a lot of that money had been taken from other planetary missions. I asked Spear what it was like, running a mission that went so smoothly, at least in space. "A project like this is a little like a family, and when things get routine, people get into squabbles about small things," he told me. "People get into rows about such trivial matters as travel money. Denver and J.P.L., which in an emergency work extremely well together, now come up with opposite ideas for solving small problems and get into fights. That's how I know the mission is going well."

One sign of a mature mission, Spear told me, is that people begin to leave. About twenty percent of the team had left for other assignments—among them Slonski, who was now running another project, the space infrared telescope facility. (When I ran into him a little later in the lobby of the SFOF, he told me worriedly that his new project had just lost $16 million in funding, and if it suffered any more cuts, maybe he'd be back on Magellan.) In another month, Spear said, he himself would be moving on to a new job—NASA likes to rotate project managers, and he was being transferred to head a new program for small projects. Increasingly, future missions would no longer be multi-instrument billion-dollar ones, such as Viking or Voyager, or even half-billion-dollar ones of the complexity of Magellan. Rather, they were apt to be dedicated to a single question, such as the detection of asteroids near Earth or whether there was water in permanently shaded craters near the poles of the moon. James Scott, the mission director, would become acting project manager of Magellan.

By November 1991, Magellan was three-quarters of the way through its second cycle—the first cycle had been completed in mid-May. Over ninety percent of the planet had been imaged—and many of the parts that had been missed in the first cycle would be filled in during the remainder of the second and third cycles. The third cycle, like the first and the second, would be devoted to mapping. J.P.L. had already announced detailed plans for the hoped-for six-cycle mission. Though there would be some limited mapping in the fourth cycle, that cycle would be devoted mainly to getting gravity measurements in the old elliptical orbit, whose periapsis would be lowered for this purpose from its present 290 kilometers to between 180 and 200 kilometers. (Mapping cannot be done at the same time as the gravity measurements, because the antenna must be facing Earth so that the DSN can track the faint changes in the Doppler shift of the carrier signal.) A period following the fourth cycle would be devoted largely to circularizing the spacecraft's orbit at 250 kilometers with the aerobraking; and the remainder of the

....

mission would be devoted to taking gravity measurements in circular orbit, which would provide far more complete coverage than the readings taken in the fourth orbit.

Whether there was money for all this was another question. When I had seen Spear, he had told me that, in Washington, planning was already under way for fiscal year 1993, which would begin in October 1992, and he and other Magellan managers were seriously worried about whether funding would last through the fourth cycle—let alone the additional time needed for aerobraking and gravity measurements in circular orbit. The other missions that had been robbed to support Magellan were sharpening their knives.

When I had seen Saunders a month earlier, at a presentation he and other Magellan scientists were giving at the National Air and Space Museum in Washington at the end of October 1991, I had asked him to fill me in on the thirty percent of the planet I had not seen at the end of the first cycle, from eastern Aphrodite eastward to Beta Regio and beyond, to the point where the imagery had begun the previous September. We were at a reception given just before the talks, on a balcony overlooking several spacecraft hanging on wires from the high ceiling. Magellan was not among them, but a test model of Voyager was—and I learned later that NASA had borrowed back some of the Voyager model's antennas and thrusters when it was doing testing for Magellan.

"The big news is Artemis Chasma and two other smaller chasms, Diana and Dali," Saunders said, grabbing a celery stalk filled with cheese dip from a

....

passing tray. "Artemis Chasma is the biggest—it's between two and three kilometers deep and 150 kilometers across. All three are arc-shaped, and the inside of the arc is raised up and then drops down. They are a dead ringer for some of our oceanic trenches on Earth which are subduction zones. Dan McKenzie is interested in this. There are no signs of spreading anywhere in Aphrodite, and if there were, there would be a problem for plate tectonics, because the rifts where spreading might occur trend north and south, and the chasms trend east and west; for plate tectonics, the rifts and the subduction zones ought to be parallel.

"A bit farther east, and north of the chasm area, there is a plain with a channel, Hildr, cut by lava, which is the longest channel ever seen anywhere in the solar system—it is 6,800 kilometers in length, 200 kilometers longer than the Nile, the longest river on Earth. Its width is constant over the entire distance. It's a mystery how lava could stay liquid all that distance." (Later, in Pasadena, I heard a talk by a team member, Vic Baker, who had studied Hildr, and who said that to cut a channel that long the lava had to have the consistency of water, which implied that it was rich in carbonates; the only way to get it would be from a plume.)

"Farther east of Hildr is Maat Mons, the scorpion's sting, which we thought was a mountain but we now know is a volcano. It is the second-highest peak on Venus, after Maxwell, but it has none of the reflective material associated with high altitudes. John Wood believes the reflective material is the result of iron in the basalt on Venus becoming weathered by sulphur dioxide in the atmosphere, which it will do at lower temperatures such as you get at higher altitudes; the iron turns into pyrrhotite, which is reflective in radar images. Some of Wood's associates think that the weathering process occurs very rapidly—within ten years—and they see this as evidence that Maat Mons is an active volcano that erupted so recently that the weathering hasn't occurred yet. But I think Wood has some reservations about this." (He did. When I heard him talk three weeks later in Pasadena, he said he believed the weathering could take millions or tens of millions of years.)

"Then we came on around to Beta Regio. One theory had been that Beta had a couple of mountains that were probably huge shield volcanoes, but it turns out that only one is, while the other is a dome. And that brought us back almost to where we started mapping."

Later, all the guests filed into a theater for IMAX films. IMAX is a technique that uses exceptionally large screens. The theater had rows of seats that angled upward steeply, keeping the audience close to the screen, which was seventy feet tall and about twice as wide; wherever you were sitting, you felt as though you were inside the picture. After the talks—given by Saunders, Pettengill, Head, and Solomon—Saunders announced that there would be a showing of two short films put together in the same way as the image of Danu Montes I had seen earlier, where a three-dimensional picture is teased out of a two-dimensional radar image by combining it with altimetry data in an elaborate computer process that allows the viewer to move around within the image as though he were aboard a spaceship. (The same technique developed for Venus would later be used by the Air Force to familiarize pilots airlifting supplies to communities within the old Soviet Union with military airfields unknown to Americans; films of the approaches were generated from satellite photography.) Then most of the speakers left the platform and took seats in the front row for a better view, but Saunders remained on the platform, behind a metal railing. For the first film, a curtain pulled partway open, revealing a small area in the center of the giant screen, the size appropriate for a small movie house. The film, with the same twentyfold vertical distortion I had seen earlier, made a spaceship-like swoop around Artemis Corona (at 2,600 kilometers in diameter, as big as Texas, and the largest such structure on Venus), passed a large whitish crater called Mitchell near one rim of the corona, and then skimmed down over the rim and into the long, curving slot of Artemis Chasma, as though the viewer were aboard one of those little fighter craft in the film *Star Wars* as it dropped into the groove encircling the Death Star. The chasm was filled with smaller grooves running parallel to each other or crisscrossing, like railroad tracks. Out of the chasm, we soared across an orange plain and dipped into two more chasms—Diana and Dali—before swooping on toward Maat Mons, towering steeply like a Mayan pyramid but with a blackish crater tipping at a jaunty angle from the top.

That trip done, a curtain pulled back, revealing the full screen—the next film had been processed for the IMAX projector. Though the screen was as big as the side of a large warehouse, not a single pixel could be seen—proof of what I had been told earlier about the vast quantity of the Magellan data and the resulting high resolution of its imagery. The spaceship, with the audience in it, hurtled over an orange plain in Alpha Regio, roared over the seven pancakes, and then zoomed around several coronae until it headed across an undulating series of lava flows— what with the IMAX on top of the vertical distortion, the whole auditorium seemed to undulate with it, making at least one viewer feel a little wobbly. Over on the right was a blip on the horizon—Maxwell Mons, too far off to be imposing. But straight ahead was Gula Mons, seemingly a jagged, round-topped peak (reminiscent of the one in *Close Encounters of the Third Kind* where the spaceship came down), whose reflective material was cascading from its top to the plain below. Long fingers of it seemed to streak tens of kilometers across the plain, leading us in a bouncy ride straight to the mountain's foot, where we stopped. The audience was wiped out, but after a moment to pull itself together, it applauded wildly. "You should have seen Saunders!" Head told me later. "He was holding on to the railings for dear life, like Captain Kirk taking the *Enterprise* through a time warp."

CYCLE 2 had begun on May 20, 1991, and though during it Magellan would make a special effort to image the south polar region largely missed by cycle

1, it was not expected to cover as much of the planet as cycle 1, because of the overheating problem and the need to do the double-hide maneuver. (During the second cycle, sixty-five percent of the surface would be imaged, more than the engineers had originally anticipated, but less than the first cycle's ninety-odd percent.) Cycle 1 imagery had been taken with the spacecraft looking to the left at an angle of approximately forty-five degrees (though it was constantly changing to improve the quality of the imagery). Because the curvature of the planet did not match the arc of the trajectory, the angle of incidence—the angle at which a viewer saw the surface in the image strips—varied from forty-five degrees at the equator to only sixteen degrees at the pole. The cycle 2 imagery was taken with the spacecraft slowly changing its attitude, so that the image strips had a constant twenty-five-degree angle of incidence for most of its mapping pass. Furthermore, the SAR was looking to the right, with the idea of seeing the backside of features that had been seen on the first cycle. This was too great a difference from the first cycle for stereo, whose viewing angles have to be from the same side and cannot be too far apart. Accordingly, on the second cycle, the spacecraft on some occasions changed its attitude back to the left, at twenty-three degrees to the surface (or about midway between the changing angles of incidence of the first cycle), for imagery which, when combined with that of the first cycle, for most places would give stereo—these exercises tested the concept of stereo with radar, which had never been done before. If the tests were successful, the third cycle would be devoted entirely to twenty-three-degree left-looking imagery for global stereo.

On the second cycle, the scientists were also on the lookout for any changes that might have occurred in the imagery of any particular place in the eight-month interval. In August, Jeffrey Plaut, Arvidson's graduate student who had wanted to remove the sand dunes from people's heads and who had just become a post-doctoral research associate at J.P.L., had found that a change had apparently occurred: a steep slope in Aphrodite imaged in November 1990 at forty-five degrees from the left appeared eight months later, when the same spot was imaged (for a stereo test) at twenty-two degrees from the left, to have collapsed in a landslide—the top of the slope was farther back and a large smudge appeared at its base, which Plaut interpreted as the debris. This was very exciting—so exciting that NASA released the news to the press—because no other signs of activity between the two cycles had appeared on any of the pictures: no eruptions, no changes in the windblown deposits on the surface. However, a few days later, after the spacecraft had switched back the other way so that the SAR was imaging the same territory at twenty-five degrees to the right, there was no landslide at all—just a reflective cliff face. The likeliest explanation for the misinterpretation was a sort of optical illusion of the radar called layover. Very likely, the image that showed the landslide was taken straight down the slope, whose angle must have been the same as the viewing angle, twenty-three degrees; because the effect of the radar is to make the back slopes of features look longer, the result in this case was to smear the image and make it look as if there were a sheet of debris. Thus, the landslide joined the concave pancakes and the flat-bottomed Cleopatra—not to mention the precipitous craggy mountains—in the album of romantic Venusian landscapes. (A couple of months later, a group of suspected sand dunes that apparently were present in second-cycle imagery but not in first-cycle imagery, but in fact were not there at all, was added to the same album.) The landslide and the dunes, like some of the other romantic views, are examples of what happens during a mission under pressure to get the data out, to interpret it, and to announce new discoveries.

The vanishing landslide was disappointing, but it by no means excluded the possibility—indeed, the probability—of landslides elsewhere on Venus, which had occurred, if not in the previous eight months, then at earlier times. During my visit, I heard a talk given by Michael C. Malin, who had recently been at Arizona State University but had set up an independent company for building space-science instruments, and who produced many pictures of what looked like landslides from previous times in many places. Though they were all over the planet, many of them were in chasms in the equatorial highlands. This is an area that is volcanically, and presumably seismically, active, and Malin believed the slides had been triggered by earthquakes—or Venusquakes. If Venus is as seismically active as Earth, Malin thought, there should be one major slide a year. If Malin is right, then if the Aphrodite slide wasn't a slide, it should have been.

In the science room, a handful of geologists were peering through stereoscopic eyepieces at pairs of enlarged F-BIDRs of the same terrain that had been taken on cycle 1 at up to forty-five degrees to the left and those taken on cycle 2 from twenty-three degrees to the left. This, of course, was real stereo, as opposed to the computer-generated stereo images which were now being produced in abundance at Flagstaff and at the image-processing laboratory at J.P.L. Real stereo was superior. The computer-generated stereo depended on combining a single image with altimeter data, which was taken every five kilometers or so and—even when it wasn't deliberately distorted— could be off in altitude by several tens of meters. Real stereo, on the other hand, was as good as the resolution of individual pixels, 500 meters or a tenfold improvement, and the altitudes had only a very small error. "It's not synthetic at all; we can measure the actual heights of real scarps!" Schaber told me when I talked to him later. The stereo tests had, in fact, worked excellently. Saunders, who frequently had to argue at budget meetings for enough money to keep Magellan operational, said that he had given the NASA managers such glowing reports of the stereo that one of them had asked why it was that Magellan was doing anything else. "Maybe I oversold the stereo," Saunders said.

Henry Moore, whom I had seen earlier at the science team's drafting and viewing tables, invited me to look at a corona in stereo. Moore is a white-haired, crew-cut astrogeologist whom I had first met when he was in charge of soil mechanics for the Viking landers on Mars, trying to gauge the strength of the

Martian soil from the electrical force necessary for Viking's scoop to dig a trench. Moore is very much a geologist's geologist. "Stereo gives us a whole new dimension," Moore said. "The sort of thing you can get is the relief of small features. You can't do this with the altimeter because its measurements are so far apart. You can see—and measure—the depths of small craters or the heights of coronae, and see their precise shape. Look at this corona!" I looked through the binocular eyepiece and saw a blur. Moore told me to adjust the two strips of imagery until they were matched for my eyes. At length I saw what looked like a fingernail on Venus. It was. Mine. A little further maneuvering, and a bit of Venus in 3-D snapped into focus. Everyone's adjustment to stereo is different, but some geologists become so adept at it, Moore told me, that they don't need the eyepieces—they just look at two adjacent strips and see stereo. The corona jumped from the surface like a fedora left on a lava flow—for a brim, it had a trench running partway around it. Moore pointed out that the top of the corona was sunken—as is the case with the tops of some fedoras, if they are pushed in. "Before we had stereo, we thought the tops of most coronae were rounded, like derby hats, but now we see that some of them are concave on top. A lot of other types of domes turn out under stereo to have slumps in the center, too. This is important for understanding their morphology and evolution." Moore moved the two strips farther down until I was looking at a plain flooded by a series of lava flows such as I had seen Moore and some graduate students trying to untangle the previous spring. Under stereo, the flows sprang into relief—it was quite clear how thick they were, and which ones lay on top of which. "Under stereo, it's much easier to build up a chronology for the flows," he said.

Two other small features that were coming under scrutiny with stereo were ticks and pancakes. I had heard that Mark Bulmer, John Guest's graduate student, had raised the question that ticks might be a degraded form of small domical features such as pancakes. Bulmer was leafing through some F-MIDRs at an adjacent table. (There were nowhere nearly as many scientists working in the science room and the cubicles as there used to be. The logjam with turning out the compact disks of the Magellan imagery had been broken, and at the time of my visit about twenty-three disks, representing about half the first cycle, had been released, and another eight would be out in a few weeks. More than ever, the scientists were working at their own laboratories. In addition to Moore and Bulmer, only Basilevsky and a young staff assistant to the science team, Gregory Michaels, were using the room much during my visit.)

I asked Bulmer why he thought the many-legged ticks were degraded domes such as pancakes. "Because ticks have steep margins and flat tops—they are dome-like in their morphology, but the edges appear to be failing," he said. "They are the same size as pancakes and other small domes—I'm talking about domes made by material extruded onto the surface, not the bigger domes that are made by lava intruded into the crust. Small, extrusive domes, such as pancakes, either drain back into the vent that made them, in which case they collapse internally, or they just degrade, with slope failures around the edges. There's a whole spectrum of these things."

(The day before, I had seen Bulmer poring over a notebook of images of ticks that had been gathered by Michaels, a trim-looking young man in his early twenties who had received his B.A. from Boston University the previous June and had then talked his way into a job with the team. "I always wanted to work here, and I'm persistent," he told me later. His job was filing images. "When I first saw pictures of the ticks, I wondered what they were, and I began keeping a special file of them," he said. "Then a scientist working here said she was interested in them, and I told her I had all the references. That's how I started." Now he was trying to classify them. Everyone on the team who was dealing with a multiplicity of a certain feature—Schaber with the craters, Bulmer and Michaels with the ticks, Phillips with regiones, and Stofan with the coronae—invariably started off counting and classifying; and just as Linnaeus's classification of plants and animals was a major tool in understanding evolution in nature, so it was with the features on Venus. Michaels told me he hoped to go on to graduate school in planetary science; he had his eye on Brown and Washington University, the haunts of Head and Arvidson, and his prospects struck me as good.)

Bulmer pulled out some of his own images and drawings—having recently arrived from England, he too had been using some of Michaels's references. One tick, evidently a onetime pancake whose lava at an early stage had drained back down the vent from whence it came, looked badly squashed, as though a rock had been dropped on it. I asked what the legs that radiated outward were. "They are the result of slope failures," he said, pulling out a picture of a particularly tick-like tick. Under a magnifying glass, the legs were promontories whose extremities represented the original rim of the tick. The material between them had sloughed away on both sides in arcs, so that the entire circumference of the tick had a regular, scalloped shape, as if a man with a jackknife had whittled a tick by carving deep crescents around the perimeter of a wooden disk.

Bulmer said, "What I've been trying to do recently is to understand the different types of slumping. One mystery is that, although you know there's been a slide, you don't see any sign of the debris at the base of the tick. It makes you look for mechanisms. It makes me think of submarine landslides—slides roar down the abyssal slopes at express-train speeds and shoot long-distance out onto the ocean floor. This happens because of the high pressure. Perhaps the avalanches from the ticks are doing the same thing under the high pressure of the atmosphere and disappearing out of sight—or spread out so thin on the plain you can see no trace of them." I mentioned that Malin had apparently found a number of slides *in situ*, but Bulmer pointed out that many of those had been in highlands, where the atmospheric pressure would be much less. (So much, once again, for the observations of a geological patzer.) The ticks, like the pancakes, were mostly on lowland plains.

．　　．　　．

IN Washington, and again at a team meeting in Pasadena, Head gave talks about what he and other members of the volcanology working group—of which he was now chairman—had been doing, in particular compiling maps of the distribution of different types of volcanoes and flows. In the division of labor with John Guest following the schism of the geology and tectonic-processes group, Head had taken over the large volcanoes and Guest had wound up with the small ones plus a few vents that had once oozed lava onto the lowland plains. (In all the geopolitical maneuverings between the geophysicists and the geologists following the spreading battle at Aphrodite, Guest, perhaps because he was six thousand miles away in England, had wound up as the big loser.) As always, there was substantial overlap, especially when it came to a catalogue of all the volcanism, large and small. The massive job of locating, describing, and classifying every volcanic feature on Venus had been done by Crumpler, Head's spreading ally, and by Crumpler's wife, Jayne C. Aubele, also a geologist at Brown. With a sort of global census now in hand, Head was able to say that eighty percent of the surface was formed by volcanism, and that the planet exhibited every type of volcanic edifice known on Earth.

The global distribution seemed fairly random—with the major exception that the central area, a huge triangular zone constituting twenty percent of the planet in back of the scorpion's tail between Beta Regio, Atla Regio, and Themis Regio, which included most of the central highlands, had about three times as much volcanism as the rest of the planet. (In his talk in Washington, Head could not resist saying that on Earth areas of increased volcanism are often where plates come apart or come together.) Alongside the volcanic-distribution map, Head projected a second map showing the global distribution of impact craters of all sizes—the work largely of Jerry Schaber and his colleagues. The craters appeared scattered all around the planet, with no great concentration anywhere. On the basis of the Magellan crater counts, which were now complete, the oldest parts of the surface were estimated at 800 million years, and the estimate for the average age was revised upward to 500 million years—only a hundred million years older than Shoemaker and Schaber had estimated in 1987, from Venera data. The estimate that volcanism produced only two cubic kilometers of lava per year, a tenth the amount produced on Earth, was based largely on the old figure for the average age; accordingly, the estimate for the annual outflow of lava was now reduced to perhaps one and a half cubic kilometers of lava per year.

On Venus and most other planets, of course, lava flows, craters, and the age of the surface in different places are closely related. Lava flows overrun craters, burying them and causing them to disappear. If the craters older than 500 million to 800 million years had been filled in by lava flows, and if craters were still being filled in that way, clearly something was wrong, because the cratering rate in the Beta–Atla–Themis triangle was the same as everywhere else, whereas, judging by the greater rate of volcanism and resurfacing there, many scientists feel it should be a third as much. "How do you accommodate the two maps?" Head asked in Washington.

There seemed to be two choices, neither altogether satisfactory. One theory, which had been suggested formally by Schaber and a colleague, Robert G. Strom of the University of Arizona at Tucson, in January 1991, when about twenty percent of the imagery of the first cycle had been processed, held that Venus had undergone one or more planet-wide resurfacing events in its history, the latest ending about 500 million years ago, and following which volcanism declined significantly (but did not cease), so that the cratering record since then has been largely preserved. It was this possibility that Schaber had had in mind in November 1990, at the time that only fifteen percent of the planet had been imaged, when he, like Saunders, had been thinking that if the craters around the rest of the planet proved to be so young in age, some basic ideas about Venus would have to be revised.

The other theory, which Phillips had put forward at about the same time as Schaber and Strom's proposal, was that the outpourings of lava over a very long period—a billion or more years—had been fairly constant, and at a rate in step with the cratering rate, so that craters were filled in in one place as fast as they were formed in another, with the result that for the most part wherever you looked at Venus the age of the surface would appear the same—500 million years old. It was in part the possibility of rebutting Phillips's theory that had reinforced Schaber during the long, unglamorous nights he put in slogging through crater counts.

As Schaber and also Saunders had suspected, the question was creating a great deal of tectonic stress among members of the science team—indeed, as much as the question of spreading had done, for it also involved basic differences in the thinking of geologists and geophysicists. At the same time, the tectonic crosscurrents resulted in a scientific tessera, so that geologists and geophysicists sometimes found themselves on the same side of the argument. For example, Head favored Phillips's steady-state theory, even though it flew in the face of the geological evidence of the cratering rate in the Beta–Atla–Themis triangle. He was strongly in Phillips's camp. Head, like many geologists and also like many geophysicists, had a built-in disinclination to believe in cataclysms —or what they think of as cataclysms. ("I naturally lean toward a uniformitarian approach, because that is how things work most of the time on other planets," Head told me.) Phillips, and also Solomon, as geophysicists were opposed to what they called the cataclysmic theory, too, because they were having difficulty thinking up a satisfactory geophysical mechanism that would explain why a planet's internal heat would shut down, as they put it, half a billion years ago.

It was precisely this basic geophysical and geological prejudice in favor of uniformitarianism, both on the surface of the planet and inside it, that Schaber had known would be challenged by the crater counts. Schaber also accused the geophysicists of misquoting him. He and Strom had never said the planet's heat would shut down; rather, they had said that the volcanism was now much less than it had been during the period of global resurfacing, but had not stopped. He had a more gradual slowdown in mind, and he eschewed the label "cataclysmic." Which of the two theories was correct was of basic importance for understanding the inner workings of Venus—an added reason, of course, why feelings ran so high, particularly among the geophysicists who found themselves having to conform their ideas of mantle dynamics to something as mundane as crater counts and volcanic distribution.

When I talked to Schaber a little later, he said, "The problem with the steady-state model is that it requires uniform or at least random volcanism almost globally over the last half billion years, and there is no known mechanism for that. However, we *can* show that the impact craters are randomly spread over the surface, and we also know there is a mechanism for it—that's the way asteroids or comets hit a planet. In particular, the random distribution holds true for the biggest craters over thirty-five kilometers in diameter, which are large enough to be unaffected by the atmosphere. If you find gaps in the distribution of the smaller craters, it may be because some of the smaller asteroids burned up in the atmosphere, but with the larger ones you know they all hit the planet, and consequently they are the best test of randomness. There are two gaps in the random distribution of the larger craters—but the gaps are inconsequential, and in both cases they are in areas where the lava flows look younger than 500 million years, so that some of the craters may have been filled in. We are not saying that the volcanism shut off entirely then (as I have to keep pointing out). However, only about ten percent of the planet has been resurfaced since that time—and that ten percent is concentrated in three areas of the Beta–Atla–Themis triangle, which we are prepared to grant the steady-state people. Incidentally, 128 of the craters are thirty-five kilometers in diameter or larger, and that percentage is about what you would expect if the average age of the surface is 500 million years. And I'm certain the smaller ones haven't been covered up, either. Though the atmosphere burns up the smaller bodies increasingly as they diminish in size, this happens at a predictable rate, so that the flux can be adjusted for Venus; and the adjusted ratios also fit the 500-million-year age, suggesting that the smaller craters have not been covered over very much. Only about four percent of the craters have been eliminated. And the clincher is that the cratering rate in the tesserae, which everyone agrees are older than the lava plains, is the same on a global scale—the impacts on the tesserae provide a control for the impacts on the plains, proving that the craters there haven't been covered up."

Because of the atmosphere, the total number of craters, now risen to 840 larger than two kilometers, is a small number for a planet the size of Venus, and they don't scatter around the planet exactly evenly, the way they would if there were more of them. There is, however, considerable agreement that they are scattered at least randomly. The randomness of the cratering on Venus was established by Roger Phillips and an associate, Richard Raubertas, a statistician at the University of Rochester. For Phillips, Raubertas ran hundreds of what scientists call Monte

Carlo tests in a computer, which mathematically threw a number of mathematical objects—in this case, the same number as the craters on Venus—at a mathematical sphere representing the planet. After each test, the computer made a number of measurements, such as how close the craters were to each other. Phillips told me, "Later we averaged all those measurements from all the tests, and when we compared them to the same measurements for craters on Venus, we found that Venus's measurements were right in the middle of all of them."

Schaber told me, "For the steady-state theory to be right, Phillips and Head would have to show that volcanism, too, has been evenly spread over the last half billion years, in order for it to leave a random pattern among the randomly spread craters, but they can't show that because it hasn't been. If it were, they would have to show why it was, and they can't do that, either, because, as I said earlier, there is no known mechanism for random global volcanism. Also, if volcanism was going on randomly all over the planet, you would expect to see a great many craters that were partially flooded. The craters that are flooded are very few in number—four percent, when, according to the minimum estimate for the steady-state theory, at least forty to sixty percent should be flooded—and most of these are in the equatorial highlands between Beta and Atla, where we know the volcanism is concentrated. (The figure of from forty percent to sixty percent came from Strom, who has been doing some Monte Carlo modeling for our side.) The observations are much more simply explained if the volcanism declined significantly 500 million years ago and what we are seeing is the random cratering since then."

Later in the fall and winter, Phillips continued to push for his steady-state theory. With respect to Schaber's argument that volcanoes weren't random, Phillips pointed out that it was not volcanoes but flows that filled craters; and although volcanoes were concentrated in a few areas and hence were not random, flows apparently covered more than eighty percent of the surface of Venus. Schaber responded that the vents from which flows flow are also concentrated in the Beta–Atla–Themis area. With respect to Schaber's argument that there should be more partially filled craters if there were random flows, Phillips pointed out that Mylitta Fluctus, that flow of flows, has obliterated all craters in its path. The small flows postulated by Phillips were on the average 400 kilometers in diameter and half a kilometer deep— enough to swamp almost any crater. Later that fall, Phillips got Raubertas to randomly fling circles representing flows of that diameter at his computer model of Venus, along with his random barrages of craters, and found that the result was startlingly like what Magellan was finding on Venus. Schaber is contemptuous of Phillips and Raubertas's disks as representing lava flows or the way they fill craters. Flows are not disks thrown at a planet, but have been observed to operate quite differently, he says. Phillips says Schaber doesn't understand the way geophysicists do modeling. "When you run a computer model, you use circular patterns for mathematical convenience,"

Phillips said. "That doesn't mean we really think the flows are circular."

The scientists took their arguments to the Lunar and Planetary Science Conference in March 1992, where Schaber and his allies (which now included Soderblom and Kirk from the U.S.G.S. at Flagstaff, Moore from the U.S.G.S. at Menlo Park, and of course Strom from the University of Arizona—a flying wedge of geologists) made what were generally considered strong presentations. Schaber told me later that he and Strom and some of his other allies won over most of their colleagues at the conference. He even thought that Head appeared to be coming around—which in fact was true. Phillips, though, demurred vigorously. Solomon and Okerson, whom I consulted as referees, said it had been a draw. Solomon felt (and so did Phillips) that neither model was right—each was what the two scientists called an end-member model, or an extreme, with the truth lying somewhere in between. Phillips thought that within the last 500 million years he could detect globally several areas of different crater densities, perhaps as few as three, which would argue for three flooding episodes 150 or so million years apart. Or, alternatively, perhaps there had been as many as ten episodes, on the average of 50 million years apart.

The cratering and resurfacing question was important, of course, because it involved the history of the planet's interior processes. If Schaber was right, the geophysicists would have to come up with a mechanism explaining why Venus had suddenly begun to lose its internal heat—something which many of them, including Phillips and Solomon, were unable to do. This inability in turn became a further weapon in the geophysicists' arsenal against the geologists' observations. The geologists, in turn, took this line of reasoning as another example of the geophysicists' tendency to put their theories and models ahead of the observational data.

Schaber joined forces with a geophysicist at McGill University in Montreal, Jafar Arkani-Hamed, who in 1984 had proposed that Venus throughout its four-and-a-half-billion-year history might have had periods of great convective heat loss from the interior, when there would have been not only massive outpourings of flows but also considerable compression and extension of the crust—maybe even something like plate tectonics; and that, because of these episodic bursts, the core would have begun to cool about a billion years ago, causing the convection to begin slowing down. (The oldest terrain, of course, was 800 million years old, and the average was 500 million.) All this, Schaber felt, fitted in not only with the cratering record but also with other observations. For example, the tesserae could be seen as the remnants of the last violent period of plate tectonics—the folding and faulting then would have obliterated all previous cratering, accounting for the fact that the cratering rates on the tesserae matched those of the lava plains, which together would have begun accumulating impacts as volcanic and tectonic activity slowed down. (There were almost as many explanations for the tesserae as there were scientists. Solomon and others, for example, thought they were caused by the compressive action of crustal movement in more than one direction; Phillips thought they were collapsed regiones; and Bindschadler thought they were made by sheet-like downwellings.)

Schaber and Arkani-Hamed are collaborating on a paper combining their geophysical and geological arguments. The two hit it off very well, despite the fact, Schaber told me, that geologists and geophysicists commonly think differently, or, as he put it, "use different sides of their brain." He feels that their collaboration shows the good things that can happen when the two sides of the brain are brought together. Schaber once asked Arkani-Hamed whether most geophysicists thought geologists were stupid. Arkani-Hamed replied, "Yes. But I teach my students the opposite. I tell them that geologists may not know much about math or physics, but they contribute observation and analysis, which is just as important."

Solomon and Phillips took a dim view of Schaber and Arkani-Hamed's alliance. Schaber and Arkani-Hamed's thinking, of course, was quite at variance with Solomon's and others' belief that Maxwell was very young and would largely collapse a few million years after the convection or plume that held it up stopped. If Schaber and Arkani-Hamed were right, many of the surface manifestations of convection would be more than 500 million years old. Their theory also flew in the face of the argon evidence, which had pushed any possible plate tectonics back far beyond that. When I asked Solomon about Arkani-Hamed's model, he said he was familiar with it, but that it had numerical difficulties not appreciated by geologists; besides, other modelers had been unable to reproduce the behavior Arkani-Hamed and Schaber described. Phillips felt that Arkani-Hamed's model had serious petrochemical problems. Solomon told me that other people were looking into theories of episodic heating, but so far he had not heard of one that he thought was convincing. Phillips told me, "You should know that geophysicists can find a model that will fit any set of observable facts." He added gloomily, "That's a joke." I knew. It was not the first time I had heard it.

Joke or no, the last laugh—or satisfied smile, at least—seemed to belong to those geologists who thought there had been a sudden decrease in volcanism. After an international symposium about Venus that was held in Pasadena in August 1992, Saunders told me, "We are converging on the idea that the rapid-tailing-off theory is a better explanation for what we are seeing than the steady-state theory." The last few times I talked with Schaber, he sounded like a cat who had eaten a canary—indeed, in view of some of the geophysicists' somewhat patronizing attitude toward geologists, like a cat who had looked at a king. Schaber attributes the change in opinion largely to Strom's Monte Carlo computations about flooded craters—and it was the low percentage of flooded craters compared to the amount predicted for the steady-state theory, Head told me later, that finally brought him around. The geologists' understanding of the geochemists' computer-modeling techniques evidently was not as poor as Phillips had thought.

At the August 1992 symposium, several papers were offered suggesting various models for episodic heating, including one by two geochemist colleagues of Head's at Brown, Paul C. Hess and E. Martin Parmentier. They proposed that, as hot material in the mantle rises, melt (consisting of the heavier material of the upper mantle) becomes part of the crust, leaving lighter material behind. This in turn forms a buoyant layer at the top of the mantle which is resistant to heat. Consequently, the interior of the planet heats up. When the lower layers of the mantle become less dense than the upper layer, the mantle overturns—causing voluminous volcanism and tectonism on the surface. The resurfacing could happen catastrophically, on a global scale, over a million years, or more slowly, in patches, so that the planet was resurfaced over, say, a hundred million years. Over the next several hundred million years, the surface would be relatively quiet. During this period, hot material would rise once again to the top of the mantle, melt once again, and be added to the crust, leaving lighter material behind at the top of the mantle—and the whole process would happen all over again.

In another paper at the conference, Head himself joined forces with Parmentier and Hess. Like Schaber, he had been intrigued with the fact that the tesserae had the same crater density (and presumably were the same age) as the flooded plains, and like Schaber he had begun to wonder whether the tesserae might not also be, as he put it, "remnants of the episodic instability of the mantle and the catastrophic heating." He and his two Brown colleagues had a different model from Schaber and Arkani-Hamed's, though.

The overturning of the mantle would not have been a simple thing. The three proposed that, as the top of the mantle became lighter than the lithosphere above, instabilities would occur at the bottom of the lithosphere. Giant blobs of material would break off and descend as downwellings on a very grand scale, which would cause great stress. As with ordinary downwellings, lithosphere would be crumpled together and piled up over broad areas, only more so —and hence the tesserae would have the same crater density as the volcanic plains, because they were formed at the same time. What else besides the tesserae Head, Hess, and Parmentier's descending blobs might have caused can only be conjectured. Very possibly the lithosphere would have been cracked—maybe into plates. I asked Head if at this point there might have been plate tectonics on Venus; he said he was not prepared to say.

When I talked with Solomon in November 1992— three months after the August symposium—he was still unconvinced about episodic heating. He was also unconvinced about the blobs dropping off the bottom of the lithosphere. Solomon felt that the similarity in the crater densities between the tesserae and the volcanic plains was better explained by the fact that the lithosphere under the tesserae highlands, being thicker, would descend deeper and thus be weakened more by heat. This in turn would cause craters on the tesserae to deform more quickly than elsewhere, eventually becoming unrecognizable. He felt that extreme deformation was a greater cause of crater loss than flooding: there are more partially deformed craters on Venus than partially flooded ones.

But Phillips, whom I talked to at about the same time, seemed to be shifting a bit. He felt that the Hess–Parmentier model was more rigorous and based more solidly on mathematics than the Arkani-Hamed model. "If episodic heating occurred, that would be a good model for it," Phillips said. "But the model does not prove that episodic heating did occur." He still believed the truth lay somewhere between what he had called the two end-member models—perhaps the three to ten separate episodes of volcanism.

The scientists and their theories, I was beginning to feel, resembled the delicate choreography of a minuet or—what might be more in tune with J.P.L.'s Western American setting—the complex weaving in and out of a particularly tumultuous square dance. Some scientists stay put, at least with respect to their ideas, tapping their feet, as it were, in time to the music (each note a data bit)—while their partners swing around them or shoot off to join other partners. Unlikely combinations sashay down the aisle together. Last year's wallflower becomes the belle of this year's ball, and vice versa. Flexibility is always at a premium —the ability to change direction to a subtle new beat. There is a premium also on the ability to detect the new beat—to perceive a new pattern in the data— before anyone else does. Science is not an art—though some may feel it resembles art in certain respects. Science may not even be a science. What it most likely is, is a very human process for arriving at the closest approximation to the truth—as elusive an ideal, per- haps, as the last dance.

THE chasms—those arc-like troughs the Washington audience had hurtled through—perplexed everyone. Artemis Chasma partly encircled Artemis Corona, though its arc had a bigger diameter than the corona—it shot off beyond the corona for several hundred kilometers. Dali and Diana were even more intriguing, for they were by themselves, apparently unconnected to any other feature, in a plain a thousand or so kilometers to the north of Artemis, near the back end of the part of Aphrodite that made the scorpion's tail. In Washington, in answer to a question at a press conference, Saunders said the chasms could very well be places where one plate was being subducted under another. "It appears that this may be taking up some of the slight movement we see at the spreading site," he had said. The trenches had a raised lip on their outer side, analogous perhaps to the bulge which Head had seen at Freyja, where one plate might be buckling down under another. Head, who was sitting a few rows ahead of me at the press conference, took note; he told me later that this was the closest Saunders had come to admitting the kind of thing he had been talking about. However, he had no intention of reviving his old spreading theory. He merely pointed out that one of the arguments against spreading and plate tectonics had been that there was no subduction on Venus; and now, although the chasms show that subduction is possible, they do not prove that the phenomenon is global. In fact, to him they looked quite local.

One person who went further than Head had ever done was Dan McKenzie, who told me firmly, "The chasms are subduction zones, like the ones at the bottom of the ocean in the East Indies. At Artemis, there are also abyssal hills and transform faults, which on Earth are associated with spreading. I don't know if there is spreading now, but there has been spreading. There is clearly plate tectonics." Coming from a man who a year earlier—when Magellan was only a quarter of the way into the first cycle—had castigated Head and others for transferring to Venus some of the ideas he had helped develop for Earth, this struck me as a remarkable development. McKenzie, who had only recently flown in from England, would not commit himself to anything more until he had had a further look at the evidence—he had no idea what the mechanics of the subduction might be.

McKenzie was not the only person belatedly beguiled by the idea of plate tectonics on Venus. While I was in Pasadena, I heard a talk given by a structural geologist from Princeton who had recently become associated with the team, John Suppe, who went even further than McKenzie. He started off by saying, "I know plate tectonics is a bad word," and then went on to argue that the entire planet was composed of two superplates which met at the equatorial highlands, which he saw as a globe-encircling rift zone. The plate that constituted the southern hemisphere showed the most evidence of extension and hence was growing, and the plate that included most of the northern hemisphere showed the most signs of compression and was shrinking. (All this, in fact, was very similar to ideas postulated by Head and Crumpler in their final pre-Magellan paper, published in *Nature*.) "I chose the word 'superplate' because I knew it would be inflammatory!" he told me when I talked to him later. "The planetary community has not dealt with an Earth-like planet before—I am a structural geologist specializing in mountain building on Earth, and the mountain ranges at Ishtar look compressed in the same way as the Himalayas, which the Indian plate has deformed much in the way the wedge of snow pushed in front of a snow plow gets crinkled. As I look at the planet, it seems to me that there is an overriding global system—not just a lot of little things going up and down here and there." He did not envisage an Earth-like situation of constant re-cycling of plates; rather, the total movement was on the order of a few hundred kilometers—in line with Solomon's figure for crustal movement.

Though in his talk to the scientists Suppe was more tentative and even diffident about his theory—he said, "I don't want to rule out spreading on Venus, or bet much on it"—some of the regulars on the team who had labored mightily to stuff the plate-tectonics genie back into the bottle, like Solomon, were clearly afraid it was getting out again. When I asked Solomon about this, he said, "First, Suppe and McKenzie are not talking about Head's ideas, which were specific to the possibility of spreading centers in Aphrodite. Suppe and McKenzie are looking at different evidence. With respect to McKenzie's idea, the question is how far you can carry the analogy of trenches on Venus and trenches on Earth. We should be receptive to his ideas, but I don't think they are very compelling. Suppe's ideas are interesting, but I think the whole concept of plates is inappropriate. As you know, I don't think the lithosphere is strong

....

enough to support the stresses plate tectonics requires over a long distance. Both Suppe and McKenzie are very original thinkers. They develop their ideas in isolation from the team—in Princeton, in Cambridge. Neither has had a chance to assimilate Magellan's global data sets. But they have a good, intelligent view from the outside, and they are worth listening to." Clearly, Solomon believed that when McKenzie and Suppe had mastered all the data they would give up their apostasy, but in January 1992, when I talked to Suppe, he had not.

There was an alternative explanation for the chasms or trenches that most of the team seemed to favor and that I found out about when Ellen Stofan was bringing me up to date on her work with coronae. Now that the entire planet had been surveyed, she was able to report that there were a total of 293 of them. They ranged in size from 75 kilometers to 2,600—the giant, Artemis. Moreover, she had completed classifying the coronae, and could make some guesses about their evolution—as she had told me earlier she would try to do. Stofan riffled through diagrams of different types of coronae: there was a class dominated by concentric fractures on top, and another with radial fractures on top. "The radial ones might be the youngest stage in the development of coronae, because the easiest way to fracture a bulge is with radial faults," she said.

She took me on through the life cycle of coronae of different sizes until she reached what she thought was the final stage. This was the class of corona that slumped at the top and had a trench around it—the fedora with its crown punched in that I had seen

....

under stereo with Moore, and whose three-dimensional topography had not been precisely known before. Presumably, when the plume that was holding up a corona died, the corona sagged in the center; and the great weight of the corona, no longer supported from below, placed an extra burden on the lithosphere that caused it, too, to sag, forming a trench. Such a trench—as Head had pointed out at Freyja—would look like a subduction zone, for flexure would cause it to have a bulge on one rim. Gerald R. Schubert, a geophysicist member of the science team from the University of California in Los Angeles, in a talk he gave a day or so later that reflected also the ideas of his associate David Sandwell of the Scripps Institution of Oceanography in La Jolla, added that in the case of Artemis, the biggest corona of all, the load and the flexure were so great that he believed the lithosphere itself had failed; the failure may also have been abetted by the great heat of the plume before it died, which would have thinned the lithosphere and weakened it. Stofan told me that in her opinion all the chasms were associated closely with coronae—Dali and Diana chasms arguably arced around the remains of ancient coronae, and Artemis Chasma's wider arc in fact enclosed not only Artemis but a second, much degraded corona. The situation where two coronae overlap, one older than the other, happens elsewhere, Stofan says, and may indicate some movement of the lithosphere over the plume that bubbled up the corona, like a much shortened version of the Hawaiian Island chain; or, alternatively, they may reflect movement of the plume under the lithosphere.

In view of Schubert's theory and of Stofan's evidence for the identification of chasms with coronae as well as for their terminal evolutionary stage, I later asked Solomon how anybody could argue that the chasms were subduction zones. "It's more complicated than that!" he said. "It may be that the lithosphere outside the corona is being subducted at those trenches, even if they are caused by loading." This, of course, would imply that the lithosphere had failed and cracked, as Schubert suggested for Artemis. In such trenches, the lithosphere from the surrounding plain might be underthrusting the lithosphere of the carona. Whatever the case, the subduction would be relatively local. But in asking the question, I had forgotten the scum—which not only made the surface seem plastic but caused any theories about the surface to be scummily flexible. The scum, of course, permitted two seemingly conflicting ideas to be operative at once, such as the theory of the horizontal movement of the lithosphere (if not exactly the spreading theory or plate tectonics) and the plume theory, or, now, loading and subduction.

But scum—physical or metaphorical—has no rules, and while it might cause two theories to drift together in one place, it might cause a theory to rip apart in another. Schubert and Sandwell's work with the chasms and the lithosphere had caused other ideas relating the interior and the exterior of the planet to come unstuck. From their work on the amount of bending of the lithosphere under the weight of the coronae, Schubert and Sandwell had gone on to make a new estimate of the thickness and elasticity—and therefore rigidity—of the lithosphere. Later I asked Schubert how he and Sandwell did all this. "We know that at the surface, at least, the lithosphere of any solid planet is elastic, not like a rubber band or a rubber ball, but like a spring, and that if it is distorted, it will bounce back," he said. "Think of how Earth's crust sinks under the weight of a continental ice sheet and then rebounds after the ice melts—that is elasticity. For a material to be elastic, it has to be cold and rigid; viscosity or plasticity are not factors. And all solid planets have an elastic surface; the only question is how deep the elasticity extends. The previous thinking was that on Venus it did not extend very far down.

"So the first thing we did was to plot a topographical profile of the corona and the chasm," Schubert went on. "Once we had the dimensions and shape of the chasm, we imagined what forces are required to bend an elastic plate, and what thickness the plate would have to be, to be bent into this particular shape. There are well-known formulas for this, and we have to use a computer to do mathematical modeling—this is how we have estimated the thickness of the elastic part of the lithosphere of Earth from the profile of ocean trenches. The elastic thickness of Venus's lithosphere at Artemis we estimate at forty kilometers." This is far thicker than anyone had thought. Other chasms give a slightly lower thickness, but by and large Shubert and Sandwell estimate that the elastic part of the lithosphere of Venus and Earth is about the same, though certain parts of Venus's lithosphere, such as at Artemis, might be thicker.

Elasticity and rigidity go together, and to suggest that Venus's lithosphere was as rigid as Earth's was something of a bombshell. Among other things, Schubert and Sandwell's work suggested that plate tectonics could not be ruled out on Venus on the grounds that its lithosphere was not sufficiently rigid to support movement over thousands of kilometers. Schubert even told me he saw nothing in the rigidity to rule out Suppe's superplates. "But the fact that the lithosphere is rigid enough so that it could support plate tectonics doesn't mean Venus actually *has* plate tectonics," Schubert told me. "It simply means that plate tectonics are not *not* happening because of insufficient rigidity, but because of something else— like, for example, the lack of an asthenosphere."

Phillips agreed with this assessment. I asked him whether Schubert's rigid lithosphere was consistent with the observed plasticity of the crust—the signs of compression and extension, of ridge belts and rifts, that came under the heading of scumminess. A rigid crust could still be stretched here and compressed there, though it might have a greater tendency to go back to its original shape if the pressure was released. "If the lithosphere is rigid to a greater depth than we thought, the direct coupling of Venus's convection to the lithosphere may still make for the untidy planet we see," he said. "And the direct coupling may be more responsible for the scumming around than any plasticity." Even if it was rigid, the lithosphere was scummier than ever, certainly with respect to defining it; even if the scum was a rigid layer, it still behaved like scum. Solomon, moreover, was not convinced that the lithosphere was as elastic as Schubert and Sandwell thought. Depending on the temperature of the crust, which is affected by the greenhouse effect of the atmosphere, viscosity, he felt, could still be a factor. What looked like scum and acted like scum might yet prove to be scum.

And many of the old problems remained the same. For example, even with a rigid lithosphere comparable in thickness to Earth's, the problem of supporting the larger features, such as Maxwell, still had to be resolved. The stronger lithosphere of course supports more weight, but not an Ishtar or a Maxwell, any more than Earth's lithosphere by itself supports the Himalayas or an Everest, which would not exist without plate tectonics.

From their new figures for the thicker lithosphere, which in turn would allow less heat to pass through it, Schubert and Sandwell had come up with a new, lower estimate for the amount of heat lost by conduction: only between twenty and thirty percent of Venus's total heat could flow out this way. As plate tectonics had been ruled out as a factor in heat loss, and as volcanism accounted for only one or two percent of the loss, only conduction was left—and Solomon pointed out to me that if one used Schubert and Sandwell's new figure for the thickness of the lithosphere, Venus's heat loss was unaccounted for by a factor of three. When I asked Schubert if he had any ideas where the missing heat flow was, he suggested that some of it might be accounted for by the largest coronae and their surrounding chasms, which in a small way might dispel heat in a Venusian version of plate tectonics. First, the lithosphere of the surrounding terrain would cool the mantle as it was subducted—just as Earth's lithosphere does; and second, new melt coming out at the center of the top of the corona was probably spreading toward the edges, removing heat. These coronae would not just look like bubbles but would in fact act like bubbles that release heat from boiling water—only, instead of bursting, these bubbles remain in place, continually resurfacing themselves, and dissipating only when the plume that made them dies down.

Solomon, though, had other ideas about the missing heat flow. The earlier estimates for conduction had been based on a thin, relatively unrigid, lithosphere. "There are two ways out of the dilemma," Solomon said. "Possibly our estimates of the amount of heat that has to flow out of Venus is wrong—our estimate, which is based on cosmic abundances in the disk from which the planets were formed, could be off by a factor of two. Or possibly the lithosphere is not absolutely elastic, in which case Schubert and Sandwell could use different numbers for viscosity in their formula, which would bring the thickness down. Maybe both possibilities are involved." Later in the winter, Solomon and several other geophysicists, putting different numbers into the formulas and models, came up with estimates for the thickness of the lithosphere's rigid zone ranging from a high of sixty kilometers to a low of twenty, and the lower estimates would bring the heat flow back up and restore the plasticity—the genuine scum.

Solomon and Phillips agreed that the jury was still

out on the question of the plasticity or rigidity of the lithosphere and the heat flow, as it was on many other questions. "We are in a state of confusion," Phillips told me when I last talked to him late in November 1992. This, of course, is the way science works most of the time—and what impels it on. Phillips had once told me it was lucky Earth had a twin, because the two could be compared and perhaps shed light on each other. Given the ambiguities and confusions with two similar planets, I thought, it was lucky in some respects that the twin planets did not have a triplet.

Head, however, would have welcomed a triplet, or a quad, or a quint, for that matter. "You need perspective on any issue—and in comparative planetology you need as much perspective as you can get," he told me when I last talked to him, also in November 1992. "If we had not had Venus, and Magellan, we would be talking about other models for the two planets than we have now. All the models that we have today are very different from what we had before." He was thinking in particular of the models for episodic heating, which have flowered as a result of Venus. Previously, others had pointed out that Earth in its early days in some respects might have been more like Venus today—and now, Head said, these possibilities would have to be stretched to include some sort of episodic-heating model for Earth in a hypothetical time in the first couple of billion years of the planet's history when plate tectonics might not have existed. "We've had a look at how Venus has operated over the last few hundred million years, but we don't know what happened in the previous ninety percent of the planet's history," Head said. "The same is true of our knowledge of Earth, where we know only about the last twenty percent of its history. At this point, it's important not to wipe our hands and walk away from these questions. It's important to see what the threads are today that lead back to the early days, both here and on Venus."

Any major new insights now would most likely have to await new evidence—and the most likely source for that would be the gravity readings, particularly the anticipated continuous ones from circular orbit, following aerobraking after the fourth cycle, which would help relate the surface features to what was going on inside the planet, map the present-day upwellings and downwellings, and possibly even shed light on the rigidity of the surface. Though there would be some gravity measurements during the fourth cycle, these would be limited in good resolution to the narrow equatorial belt beneath periapsis. The most comprehensive measurements would be those from circular orbit, following aerobraking. If the spacecraft and the funding held out, there might then be two more complete cycles for the gravity measurements.

THE calm at J.P.L. was shattered in January 1992, when Magellan suffered two blows that came close to knocking it out of the sky. One was electronic and the other had to do with funding, and of the two, the electronic one was the least drastic—it was the sort of problem that the engineers were good at dealing with. On January 4, when the spacecraft turned away from Earth to do its star scan at apoapsis of its 3,880th orbit, it was working perfectly; but when it turned back to Earth afterward, it was not transmitting data. It was transmitting only the carrier signal, the steady whoosh that is broadcast whenever the transmitter is on—the subcarrier whose modulations conveyed the data was gone. Clearly, the problem was not in the transmitter but in the larger communications system, the transponder, which includes the modulator, the amplifier, and other components. The spacecraft was using transponder A; it had not used transponder B since it had developed a whistle in its X-band transmission the preceding March. As they had been unable to think of a fix, the flight controllers considered they were single-string with the transponders.

The flight controllers commanded the spacecraft to switch to transponder B, the broken string. There was no whistle—but after twenty-five minutes, when the transponder had warmed up, the signal began to decrease, which meant the whistle was going to start; the whistle uses power, and hence the decrease in the signal. The flight controllers turned the transponder off and thought for two days. Then they tried to trick the whistle by allowing the transponder to warm up a bit before they turned on the transmission. But the whistle didn't fool easily. When they turned the transmission on, the whistle began once the transponder's temperature reached 35° Celsius, and as the temperature increased, it got louder and louder. It made data transmission next to impossible. The engineers tried to fool the whistle by moving the

wouldn't follow, but it did. On January 14, the
spacecraft stopped mapping and was turned away
from Earth for a planned ten days of orienting its
solar panels toward the sun to recharge its batteries.
Transponder B was left running, and on January 24,
when mapping resumed and the spacecraft turned
back to Earth, the whistle was still there. At last, the
engineers succeeded in losing the whistle—but at a
price. They lowered the frequency to a point where
the whistle did not follow; but the lower frequency
meant they had to cut the transmission rate to only
forty-three percent of what it was. That meant that
each noodle would be 6,450 kilometers long instead
of 15,000. As a result, the third-cycle imagery, instead
of being continuous (though with time taken out
occasionally to fill in holes in previous cycles), would
now have to be targeted, with the scientists selecting
the most important sites for stereo. However, the
transmission was clear, and the flight controllers
intended to keep it that way, for they were now once
again single-string with the transponders. The gravity
experiment would not be affected because it uses the
carrier signal only, and Magellan was double-string
with that.

THOUGH the engineers could save the mission
from major transponder or computer problems out
around Venus, they had far greater difficulty with
the financial problems in Washington. In the early
fall of 1992, NASA headquarters submitted its budget

requests for fiscal year 1993 to the Office of Management and Budget, which had marked it up (Washington talk for making its comments) and returned
it. The NASA budget passed back and forth between
NASA and the O.M.B. When President Bush had
presented the 1993 budget to Congress on January
29, 1992, there had been no funding for Magellan
for the 1993 fiscal year, which began on October 1,
1992, except for a paltry $1,565,000, which rep-
resented the final award money owed to Martin
Marietta. As Spear had feared, the budget—which
contained the combined judgment of NASA, the
O.M.B., and the White House—contained not a
penny for operating the spacecraft after it was only
six weeks into the fourth cycle. Nonetheless, in a
letter of instruction dated January 29 and signed by
Lennard A. Fisk, the associate administrator for science and technology, NASA headquarters told J.P.L.
to complete the fourth cycle—even though no money
was provided for it. As Okerson explained this rather
Delphic instruction to me, it meant that the Magellan
project had to save enough money from its 1992
budget, $45,900,000, to operate the spacecraft and
otherwise support the fourth-cycle gravity measurements in the present highly elliptical orbit in fiscal
year 1993. As Okerson interpreted the letter—and
interpreting the letter to the Magellan project was his
job—the third-cycle mapping for stereo would probably have to be turned off before it was complete, for
savings from the third cycle would have to be used
to pay for the fourth. This upset him. "If we don't
continue mapping until the end of the third cycle,
the spacecraft will be an empty, useless hulk for

several months," he said. Of course, there would be no money for aerobraking or for the later cycles of gravity measurements in circular orbit.

The funding problem existed largely because there were too many other planetary missions competing for the limited funds earmarked for them. In the proposed 1993 budget, there were other casualties as well as Magellan—in particular, a mission called CRAF, which stood for comet rendezvous-asteroid flyby, which was to go in tandem with Cassini. This increasingly is the way of life in planetary exploration—in 1992, for example, other missions had to be cut into to keep Magellan going. Fisk, who had made the decision to terminate Magellan at the end of the fourth cycle, told a reporter for *Aviation Week* and *Space Technology* in early February 1992 that a trend was developing toward cutting spacecraft data-acquisition budgets once the flights have completed their primary missions, to provide more funding for new spacecraft. And Wesley Huntress, who serves under Fisk as the head of NASA's solar-system exploration division, told me, "The problem is the tight budget at NASA. There's no place I can go to get money for Magellan. Killing Magellan does not make me happy. It sets a bad precedent to turn off a working spacecraft."

In the press, NASA officials pointed out that Magellan had already achieved—indeed, exceeded—its primary mission, which was to map Venus with radar one time, covering seventy percent of its surface. For a planetary spacecraft, Phillips told me when I spoke to him, "primary mission" is a bureaucratic term useful to accountants but virtually meaningless in terms of science or exploration—as in the case of Voyager, which went on to Uranus and Neptune after its primary mission was complete. A working space-craft orbiting another planet and capable of returning useful data has never been turned off before. Normally, planetary spacecraft live to a ripe old age and are treated with tender loving care, not only because of the great cost of building and launching them, but also in recognition of the near-insurmountable odds against getting them to their destination and of their almost miraculous survival in these harsh and hostile environments. To some, every additional scrap of information that can be squeezed from a mission seems an extra dividend, worth its weight in gold.

All the scientists, exhibiting a rare unity, were beside themselves. They were particularly upset about the loss of the aerobraking and the extra couple of cycles for gravity measurements, which could be done with a very limited number of people and would cost only about $10 million or $15 million—peanuts, as the scientists saw it, compared to the amount already spent on Magellan, let alone the rest of NASA's budget. In return for the small investment, the results of the Magellan mission would be vastly increased. Jerry Schaber said, "It's a waste of the taxpayers' money not to get the most out of a mission, considering the amount invested in it." Gerald Schubert said, "It's unbelievable that anyone would allow a spacecraft to be turned off before it has completed essential measurements." And Phillips said, "It's incredibly shortsighted! Magellan has a lot more it could do—we could get a great deal more out of this mission for very little money. If we learned the global convection pattern of Venus from the gravity data, we would not only know better how the planet worked but we would be much more able to compare Venus and Earth. It is that knowledge we are in danger of losing."

Gordon Pettengill, the chief scientist for the radar, whom I spoke to a little later, told me he and some of his colleagues intended to lobby Congress. As most of the scientists—unlike the engineers and managers—were not government employees, they were free to do this. They also knew the arguments better than anyone else. In other ways, too, the Magellan scientists, engineers, and managers were not without resources, and they embarked on a series of steps every bit as devious as the ones the computer technicians had employed to save the spacecraft from runaway program executions, to prevent this ultimate LOS (a program execution in another sense). The people who had lived through the RPEs in space were not going to be deterred by a little thing like a walkabout in Washington. There was one basic rule of the game: the Magellan scientists and managers could not overtly act as though they planned to do aerobraking when they had been told they couldn't—anything short of that, though, was within bounds. (The scumminess of Venus had nothing on the Magellan team, when it came to bending but not breaking principles.) An underlying reality of the game was that if they kept the spacecraft alive and maintained at least a skeleton crew of scientists and engineers, funding might be found somehow to continue the mission; there was always next year's budget.

Of course, they had to comply with Fisk's letter of

HENRY S.F. COOPER, JR.

January 29, 1992. Okerson, who had always struck me as very pragmatic and matter-of-fact, told me the day after the decision, "My job until yesterday was to argue valiantly for the Magellan mission. Today I have to figure out how to get the most out of the mission under the new rules. Very soon we are going to have to start throwing stuff off the sleigh—anything not directly related to operating the spacecraft and collecting data will be jettisoned, to save money for the fourth-cycle gravity mission. Secretaries, engineers, and scientists will have to go. The criterion for picking who goes has nothing to do with anyone's abilities or anyone's past contribution to the mission, however great. It will be based solely on whether they now contribute to mapping or gravity, or keep the spacecraft going for those purposes." All these departures—some mandatory, some voluntary —were also carried out in such a way as to keep the option open for aerobraking.

(In most cases, the people who left Magellan stayed at J.P.L. and were assigned to other missions or hoped to be. Spear, now in his new job managing small missions, told me he hoped to find new berths for many of the Magellan engineers on some of his fleet of little spaceships headed for the moon, for an asteroid, and for Mars. J.P.L. has always reminded me of the intergalactic pub in the film *Star Wars*, a way station for extraterrestrials in transit to remote worlds; for the J.P.L. engineers, when they are through with one mission, are continually awaiting transport to another planet—and usually they get it.)

It appeared that the lower down NASA's chain of command, the more anxious people were to save the

Magellan mission. Okerson, for example, was more sympathetic to the scientists and engineers than his superior, Huntress, who in turn was more sympathetic than the obdurate Fisk. When in July 1992 I saw Huntress, a handsomely bearded man, in his office in Washington, he told me that there was no doubt he would like to do aerobraking.

During the winter and spring of 1992, the visits of the scientists to Capitol Hill began having their effect; the phones in the solar-system division began ringing with calls from members of Congress or their staffs. Repeatedly, Head and Phillips were among the most successful lobbyists.

In the late winter, Huntress sent a request to the Magellan project for proposals for a number of different options. One of the proposals Huntress requested included the possibility of continuing the radar experiment throughout cycle 3, plus doing the aerobraking at the end of cycle 4, and no more. Another proposal built on the previous one but went on to ways and means of doing gravity measurements in two additional cycles. Saunders worked up a proposal for doing the aerobraking and keeping the gravity experiment going beyond the fourth mission with a crew of only fifty people—at a cost of $12 million a year.

At a meeting in early summer, the scientists voted to ask NASA for an additional $10 million in the 1993 budget to add mapping to cycle 4. If they had asked for the money to do aerobraking, the request would have been turned down, because, officially, aerobraking was not a possibility. However, if they got the money, the intent was to use it, not for radar, but for that other purpose—a plan that was not as dubious as it sounded, because once the project received an appropriation for whatever purpose, it could legally use it any way it wanted. Besides, people in the solar-system division were hardly unaware of what was going on—indeed, seemed to be abetting it.

In fact, the division had recently written the Magellan project a memo instructing it to keep open the option of the forbidden aerobraking. As it happened, many of the engineers who had been planning aerobraking were still around—though it was hard to deduce this from looking at a personnel sheet, because officially they were assigned to other tasks. In the last year and a half, working often on their own time in the manner of many NASA engineers (and scientists, too), they had refined their calculations over and over again. As a result, they had made some changes in the aerobraking scenario. For example, when computer modeling showed that hitting the atmosphere antenna-first would destabilize the spacecraft and that the antenna didn't provide much protection against frictional heating anyway, they decided the spacecraft should travel with the umbrella trailing behind, where it would serve as a sort of keel. The engineers received fresh data about Venus's upper atmosphere in October 1992 when Pioneer–Venus, almost fourteen years after arriving in orbit, finally plunged into it, vaporizing—the ultimate in aerobraking. Pioneer–Venus's fiery demise also had a public-relations value because it focused attention on the Magellan project's hopes.

But in the fall of 1992 things did not look so bright.

Fisk told Huntress that he should stop thinking in terms of aerobraking. A little later, Daniel S. Goldin, who had been named administrator of NASA the previous March and whose views on Magellan funding were not yet known, was reported to have said that the days when NASA would expect to get a hundred percent out of a mission were over—thus reinforcing an earlier statement by Fisk. The scientists had given up on going to Congress. The amount of money needed to keep Magellan running now was so small, $10 million or $15 million, that the scientists journeying to the Hill were being told, "Don't bother me for such a small sum! NASA should be able to find the money in its own budget." If it could, it wasn't giving the money to Magellan. The scientists themselves were out of money—after October 1, 1992, the beginning of fiscal year 1993, there were no funds to pay them. They kept working anyway; and in November Huntress managed to scrape up $3 million to keep them going.

In November, when I talked for the last time with Saunders, things were looking up—or at least halfway up. Huntress had asked the Magellan project to suggest a suitably interesting finale to the mission, to be called the termination experiment, after the fourth cycle ended in the middle of May. The scientists and engineers proposed doing ninety days of aerobraking, out of the 110 days required to circularize the orbit. Clearly, the Magellan team has a different definition of the word "termination" from the one in Webster's (". . . end . . . conclusion . . ."). The new orbit after this version of aerobraking, Okerson told me in late November, would be around 300 kilometers above the surface—higher than the scientists originally had wanted; but even if it was slightly elliptical, rising to an apoapsis of 500 kilometers, no one would complain.

Whether there would be aerobraking, or whether there would be gravity measurements afterward, was left an open question—though when I last tuned in, in March 1993, at least forty days of aerobraking seemed assured. Saunders told me, "We will save NASA yet from the embarrassment of shutting off a spacecraft that could do more work—in spite of itself!"

But there was more to the great lengths the scientists and engineers were going to than saving NASA's chestnuts, I suspected. To the people who had struggled to keep Magellan going through all its difficulties, and who were still arguing intensely over the data that were still coming in, ending its mission when there was more information it could collect was like destroying something alive.

(continued from front flap)

lights the continued importance of NASA space missions and the tenuous politics of their administration. An evocative narrative of the people who do science and of the challenges that confront them, *The Evening Star* is an illuminating portrait not only of Venus's character but of Earth's as well, and of the place of the two siblings in the family of planets.

HENRY S.F. COOPER, JR., has been a staff writer for *The New Yorker* for over twenty-five years. He is the author of *Thirteen: The Flight That Failed*, *A House in Space*, and *The Search for Life on Mars*. He lives in New York City.

FREYJA
MONTES

ISHTAR

LAKSHMI

Colette Patera

PLANUM

Sacajawea
Patera

MAXWELL
MONTES

FORTUNA

Cleopatra
Patera

PLA

LAVINIA PLA

LADA T

POST-MAGELLAN